Running Your Machines with SPC

A Shop Floor Guide for Manufacturing

By James C. Abbott

The Walkabout® Series

Robert
Houston
Smith
Publishers

ROBERT HOUSTON SMITH PUBLISHERS
PO Box 2193
Easley, SC 29641-2193
888-747-6484 (telephone)
888-747-3291 (facsimile)
Info@RHSPublishers.com
www.RHSPublishers.com

For information about the **Walkabout® Method**, contact
Abbott Associates Incorporated / PO Box 26177 / Greenville, SC 29616
www.AbbottWalkabout.com

Cover design and illustration by Lightbourne, copyright 2000
Content Advisor - Ron Parker
Production Manager - Kelly Boland
Graphic Design - Krista Spivey & Charles Hughey

Publisher's Cataloging-in-Publication
(Provided by Quality Books, Inc.)

Abbott, James C.
Running your machines with SPC : a shop floor guide for manufacturing / by James C. Abbott. -- 1st ed.
p. cm. -- (Walkabout series)
Includes index.
ISBN 1-887355-05-7
LCCN 00-192408

1. Manufacturing processes. 2. Process control--Statistical methods. 3. Production planning.
I. Title. II. Series.

TS183.A23 2001 670
QBI00-901888

This book is dedicated to my wife Susan. Her steady hand in raising our children and running our household has allowed me to chase my dreams.

Sue, you made this book and all of the others possible – thank you.

Table Of Contents

Preface

This book is one part of the Walkabout® Series. The series is composed of:

- *Running Your Machines with SPC: A Shop Floor Guide for Manufacturing*
- *Optimize your Operation: Stories, Tools, and Lessons for Using the Principles of Process Management to Improve Your Quality*
- *SPC: Practical Understanding of Capability by Implementing Statistical Process Control, third edition*

Organizations are changing dramatically; the last two decades have seen the pace of this change increase. Many companies are attempting to organize and run their business like it was 30 years ago, but new and more effective tools are required to keep up. These tools apply to both manufacturing and service organizations.

We view running a business, or an improvement process, like an expedition ascending a mountain to its summit. A mountain climbing expedition always meets roadblocks. The expedition requires certain essential elements to overcome these obstacles and so does the improvement process. The mountain climbing expedition's essentials include having the right equipment, access to the essentials of life (food, water, air), having a stable launch point, and properly trained team members. These essentials are referred to as the base-camp for a mountain climbing expedition, and in our improvement process, a base-camp will also be required.

For improvement to occur, a base-camp of product and process knowledge must be built. This base-camp will include knowledge about each machine's strategic correctness – its correct, consistent, and capable operation. A medium for communicating this knowledge is key. This medium must present the knowledge in a useable fashion – one that is controllable and maintainable. The base-camp must be built in a predefined sequence. The first step is "correct," followed by "consistent," and finally "capable." These three functions must be done for every aspect of the business. This includes determining the correct, consistent, and capable running of each individual machine. Finally, the product and process knowledge must be used, and the brilliant execution of our knowledge base will lead to improvements.

Now, let's look at each step in its prescribed order.

Strategic Correctness

The first step in our journey to build the base-camp establishes correctness. This correctness is divided into strategic correctness and tactical correctness.

First, the strategic correctness must be established. A correct, functional organization with clearly defined roles must be determined to cover marketing and sales, product design, process design, manufacturing, and quality assurance. The product and process factorial families must be correctly calculated. A clearly defined product and a detailed method must be created.

Tactical Correctness

The tactical correctness is the optimization of the methods for each machine to run the product. This correct method must be communicated, and our tactical workforce must perfectly execute the plan.

Optimize Your Operation explains the "what" and "how" of tactical correctness. It explains the details and subtleties of the Principles of Process Management. For tactical execution and optimization, the Walkabout® Dependency Diagram and Walkabout® Method are explained. These give a clear plan and structure for the optimization of your facility. This book explains the method for optimizing each machine to achieve tactical correctness.

Consistency and Capability

The book *SPC, third edition* deals with the latter two components of the base-camp: the consistent running of each machine, and an evaluation of the machine's capability. *SPC* explains the mathematics, theories, and implementation of the two Statistical Process Control tools (control charts and capability studies). The correct organization and operation of the facility radically influences these topics. *SPC* gives a comprehensive explanation of both the theory and practical uses of SPC.

The second step toward our base-camp means monitoring each metric for every machine to determine how consistent each product and process metric is. This consistency provides a way of predicting what will be produced time after time.

Finally, with the steps of correctly and consistently run machines or processes we are positioned to assess how well the product we make meets the customer's expectation.

The Walkabout® Base-camp

Strategic and tactical **correctness**, **consistency**, and **capability** are the three Cs that form the Walkabout® Base-camp. Just like an expedition base-camp, the Walkabout® Base-camp provides the essential elements for an effective organization. This effective organization will see immediate results that include better products, lower cost, and quicker delivery. The Walkabout® Base-camp also provides the knowledge and structure to make, sustain, and share future improvements.

The book *Running Your Machines with SPC* explains how to use the three Cs of the base-camp and the two Statistical Process Control tools. Stories and examples are used to move past the mathematics of SPC straight to the key ingredient of correctly using SPC to run your machines.

A Word about "Good Intentions" and the road to Quality Hell

All stories and examples that are discussed in my books are true. Only the names, products, times, and places have been changed to protect the innocent. As we cover the material, I will show you that many well-intentioned efforts wind up causing exactly the reverse of the result we want. Often they not only stand in the way of improvement but also cause our efforts to deteriorate. These are like golden bricks: the golden bricks of good intentions that we feel are going to help but actually do the reverse.

In the Walkabout® Series, we will discuss the idea of good intentions. Everyone has heard the saying *"the road to hell is paved with good intentions."* The golden bricks that I will describe in this book are paving a road that leads to a horrible fiery place – a place called Quality Hell.

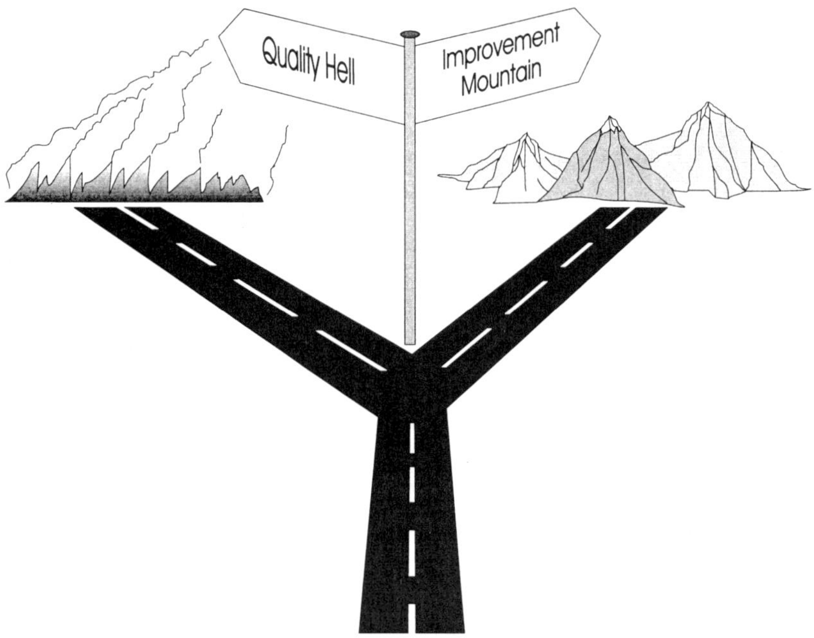

Many times throughout these books, we will be at a crossroad. One path at this crossroads is paved with the golden bricks of good intentions leading to Quality Hell. The other road is paved with the bricks of Process Management Principles and leads to productivity and quality improvement. We must assure that we have a clear picture of the correct road so that we can plot a course to productivity and improvement.

Quality Hell

Good intentions that put bricks on the road to Quality Hell (flawed logic and questionable input to decisions) will be indicated with the Quality Hell icon to the left. We will also provide practical examples of the good intentions, along with some sound common sense. Be sure to beware of the golden bricks of good intentions that we will be pointing out.

Topics that are of particular importance will be identified with the Walkabout® man symbol. When you see him, you know he is leading you up the path to Improvement Mountain.

The Walkabout® Base-camp provides a clear picture of what correct, consistent, and capable products and processes should look like. The base-camp will act as our launching point for all improvement efforts and new product introductions. It will guarantee quicker launch times, improved products, reduced cost, and faster delivery.

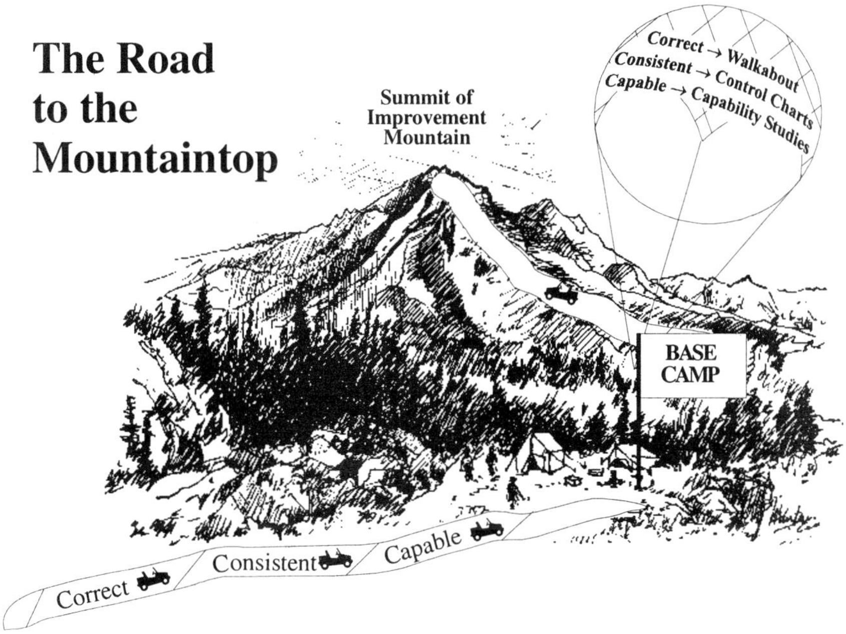

Chapter One:
Statistical Process Control

Statistical Process Control (SPC) is used to analyze and monitor a continuous process. SPC should be used to monitor both product and process characteristics of all continuous operations.

Since SPC is designed to monitor a continuous process, an ongoing surveillance program is required. The flowchart to the left depicts the steps required for this monitoring system. First, we must gather data on a regular and consistent basis. Second, this data is measured for consistency with older data using a control chart. Once a state of statistical control is established, we must begin the third step of assessing how well the data meets the customer's needs using a tool called a capability study. The final step is maintaining control with a capable process.

Each one of these steps will be explained in this chapter.

Statistical Process Control is comprised of two tools: 1) **control charts** and 2) **capability studies**.

A control chart is designed as an alarm system to detect change in an operation. Change can be deterioration or improvement. Because it is the operator's tool for monitoring a particular machine or system during its time of function, it includes an alarm system for detecting any type of shift or change. This alarm must be dynamically recalibrated to reflect the current status. The $\bar{X}$ **and R** and **Individuals** control charts are most commonly seen in a typical operation and will be the focus charts for this book.

A capability study is the tool for management to assess how well the product being produced matches the product the customer desires.

The SPC team includes control charts for tactical purposes and capability studies for strategic work. Therefore, both must work together to form a cohesive team. Let's examine each tool independently to understand how the SPC team of control charts and capability studies works together.

MONITORING A CONTINUOUS PROCESS

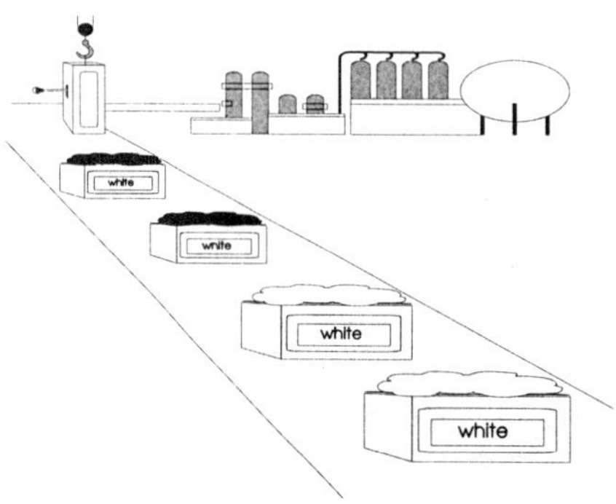

Monitoring the Candy Plant Packaging

Notice in the drawing to the left that we are packaging our candy in a white container. The third product shown is not white but has changed to black, and all subsequent product is black.

Analysis of a particular machine, process, product, system, or entity must be done in the chronological sequence. If we lose the integrity of the time of an event, we have lost a tremendous source of data.

This knowledge of when the candy package changed is critical to our analysis. The quicker change is detected, the better. The longer the package change goes undetected the more the bad product builds. We must not wait until many bad candy packages are produced. If we had lost the time of the product change, we might have mistakenly determined that the change was a random event rather than a systematic change. Instead, we must have a continuous surveillance to detect change as early as possible.

TIME SEQUENCE

In order to detect patterns, trends, and shifts, we must not only document the time of occurrence but also group the values in chronological sequence to detect the shifts. Chronological sequence is necessary for identifying causes.

If we lumped both the black and white containers together, we would have an incorrect impression of the quality level of the product. The result would be very camouflaged and confused. No decision from this kind of analysis would be valid. To get a complete production picture, any change must be separated and assessed.

Time-series

By maintaining the integrity of the time sequence, we also have the ability to take care of our time-series grouping. The time-series grouping is resolved by using control charts.

SPC OBJECTIVE

The SPC objective is much more than simply detecting a defect. Rather, the SPC objective is to monitor for both improvements and deterioration allowing us to get the most from our facilities. Like the United States army says, "Be all you can be." We want to use SPC to get all we can out of our facilities. This objective uses the Second Principle of Process Management to make good decisions. The next section is a brief overview of this principle.

SECOND PRINCIPLE OF PROCESS MANAGEMENT

Division of Labor is the framework for all aspects of decision making. It must be clearly understood to separate the strategic and tactical decisions. Operations makes the tactical decisions of running the facility. Management makes the strategic decisions of assessing the facility's suitability for the job.

The Second Principle of Process Management Summarized: Division of Labor drives the decision-making process. It dictates who will be in charge of the different kinds of decisions (strategic or tactical). We are doomed to a state of Quality Hell if the managers of the future fail to fully grasp Division of Labor.

If we have a clear understanding of the second Principle of Process Management, or this "Division of Labor", we will possess the key to appreciating the difference between tactical and strategic decisions.

Division of Labor requires a clear understanding of the differences between management/customer/sales (strategic) issues and operational (tactical) issues.

Control is a major tactical issue associated with operational duties. Capability is a major strategic issue associated with management, customer, and sales concerns.

Tactics: The art or skill of employing available means to accomplish an end.

Operational decisions are tactical and detect change.

Strategy: The art of devising or employing plans or stratagems toward a goal.

Management decisions are strategic and assess impact.

Division of Labor drives and controls goal setting and decision-making. Clarity of purpose is essential to effective goals and decisions. Operations must be tactically driven while management/sales/customer must be strategically driven.

The principle of Division of Labor will have a profound effect on decision-making. The concepts that follow must be clearly understood, for without them, no improvement is possible. These concepts are given in-depth coverage in the book *Optimize Your Operation*.

Second Principle of Process Management

Division of Labor

↓

Accountability

	Operations	**Management/Sales/Customer**
Accountability	The correct and consistent running of the facility.	Finding customers who will buy our products.
Responsibility	Running the provided facility correctly and consistently.	Providing the facility that will produce what the customer wants.
Missions	Control	Capability • Customer Specifications • Targets • Goals • Requirements
Decisions	Tactical	Strategic

Strategy comes from the Greek word strategos, which means "the art of the general." Generals make several kinds of strategic decisions including which resources to use, when to use the resources, and where to use the resources in the conduct of battle. Strategy, for our context, is the planning and provision of the correct resources, at the correct time, in the correct place.

Tactics is the execution of the strategic plan that the general has developed. Our execution of the strategic plan must be the best and most effective use of the provided resources.

For the most effective and efficient operations, we must understand our role according to Division of Labor and then execute brilliantly.

ACCOUNTABILITY

To be held accountable, we must first know whether we are in a tactical area or a strategic area. Tactical personnel should be held accountable for issues and decisions that deal with the execution of the plan. In a continuous operation, the plan deals with the facility that has been provided. The operator, who is the tactician, should be held accountable for the correct and consistent running of the provided machine.

Strategic personnel should be held accountable for developing a plan and building a facility to meet the company's needs. This facility must produce the product that the customer wants.

The book *Optimize Your Operation* contains a more detailed explanation of the Division of Labor.

VIOLATIONS OF THE SECOND PRINCIPLE

Second Principle violations must be understood and avoided in order to effectively make decisions. The kinds of situations that we will encounter if we violate the Second Principle are:

Division of Labor Violation 1 - No Thought to Capability

If we violate the Second Principle by confusing the responsibilities of management with those of operations, we have no basis for assessing the capability of the facility. We in management can only ask operations to do what the facility is capable of doing. In this way we can hold operations accountable for the correct and consistent running of the facility. Management must provide operations with a facility that is capable of producing the customer's product.

Division of Labor Violation 2 - Forced Capability

Good Trooper Award: If we violate the Second Principle by trying to place the responsibility for strategic decisions on operations, we will not be able to meet the customer's demands, and we also make a more defective product than was necessary in the first place. The harder a good associate works with improper instructions (violating the Second Principle), the poorer the quality. Quality will continue to degenerate.

Division of Labor Violation 3 - Poor Productivity

If we violate the Second Principle on an extremely capable process, we will make the machine appear less capable than it really is. Instead, we must run the machine to its own personality as correctly and consistently as possible. We must not be lulled into accepting products that are only as good as the customer wants and allow the process to float inside the customer's specification.

We must guard against any violation of the Second Principle – Division of Labor. A good associate who is put in the situation of having to run a facility with improper directions that violate the Second Principle is doomed to disaster. The better and more conscientious the employee, the worse the situation will become because the person will go to any lengths to carry out those improper directions.

TACTICAL TOOLS

Because of Division of Labor, SPC is comprised of two tools. Each is specifically designed to support either tactical or strategic decisions. The tactical tool is called a control chart while the strategic tool is called a capability study. Let's now focus our attention on the operator's role, and provide him with analytical support to best run his machine. This effort is referred to as control.

Where are we?? Before any improvements can be made or problems can be solved, knowing where we are is essential. Statistical Process Control is the set of tools that provides a clear idea of where we are in relation to our target. Before an archer can hit the bull's-eye, he must know where the target is. He must also know where his arrows are falling so that any adjustments are made in the appropriate direction.

The details of archery are essential to hitting the bull's-eye. Details are also essential to a continuous operation. Now let's focus on the first detail of the correct and consistent running of the operation.

What does Control mean?? When we say that we are **in control**, we are only talking about operational issues. The term **in control** means that we are running the facility correctly and consistently. This gives us a way of knowing where we are.

We must have tools available to us to measure a facility's uniquely correct and consistent running parameters. Control charts were designed expressly to do the job of evaluating the correct and consistent running parameters of a machine. A control chart also functions as an alarm system to alert us to changes and deviations from these optimum parameters. Remember that changes can take the form of either improvements or deterioration of the process. However, control charts cannot identify whether the change is an improvement or deterioration. A control chart's mission is very specialized: to identify change only.

OPERATIONAL DECISIONS

When we understand the concept of control, we realize that the objective of operations is the correct and consistent running of the currently available and existing facility. Operations must keep a focus on the tactical needs by making tactical decisions and conducting tactical analysis.

Tactical decisions are defined as decisions employing the available means to accomplish an end. Operations makes tactical decisions.

Operations => Tactics: Operations uses tactical decisions to better understand the facility. With this knowledge, operations can use tactics to continuously run the facility correctly and consistently.

THE OPERATOR'S ROLE

The operator must focus his attention on the provided machine. The operator must run the machine as correctly and consistently as possible with the current state of knowledge. Running the machine to the maximum is the operator's responsibility for which he must be held accountable. Tools to assist the operator must be provided with the clear mission of supporting the correct and consistent running of the machine.

The objective of correctly and consistently running the machine means getting the most out of the machine no matter what the customer's specifications require. Our tools to support the operator will thus be independent of the customer and his requirements.

TOOLS TO SUPPORT THE OPERATION OF A CONTINUOUS PROCESS

There are many tactical analytical tools that are designed specifically to support operations. One tool, called a control chart, is used on continuous operations to detect change. Control charts provide the mechanism to determine whether or not a particular machine is running consistently, and are also a means of monitoring and detecting homogeneous groups. They could also be called "consistency charts," "change detectors," or "cause detectors."

Control Chart

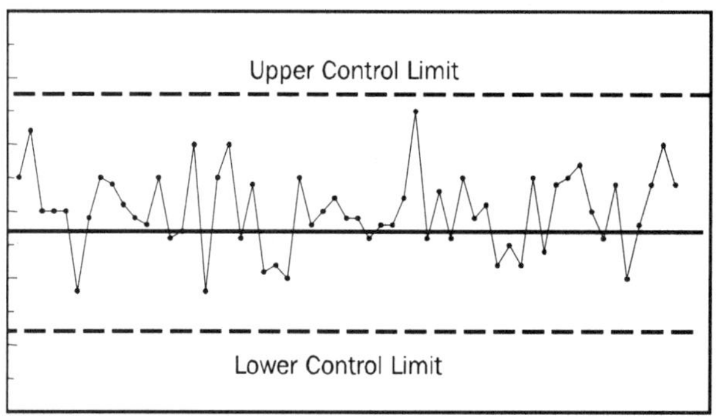

Many people attempt to make a control chart into both a management and an operations report by including specifications, goals, customer expectations, and tolerances on them. Control charts are operational tools, not management reports. Since specifications are strategic in nature, they NEVER appear on a control chart. The whole mathematical concept behind control charts assumes we are only discussing tactical issues.

Control charts were never intended to be used as an indication of how well the product fits the customer's needs. Control charts are for the tactical decision of consistently running the facility and for detecting change.

WHAT EXACTLY IS A CONTROL CHART

The objective of a control chart is to help an operator run a machine correctly and consistently (without change). This is referred to as being in control. Because change can be either improvement or deterioration, any change must be identified immediately so that management can assess the impact to the customer.

Since the change must be detected quickly, a change detector is critical. That is exactly what a control chart is.

I hope you have a smoke detector in your house. The smoke detector sits quietly in the background until smoke or fire is detected. When the detector senses these situations, it sounds the call to arms loudly and aggressively. No action is taken based on the alarm alone, but immediate inquiry into the situation is required. Once the smoke detector alarm is activated, we immediately search the house to find the cause of the alarm. When the fire's cause is identified, we can make a decision about what to do. We must know the cause of the fire before taking action because if we throw water on a grease fire, the fire will spread. This is exactly how a control chart works - instantly alerting the operator to any change and triggering our research as to the cause.

Control Consistency Alarm Chart System
Operator's Tool
1 Particular Entity

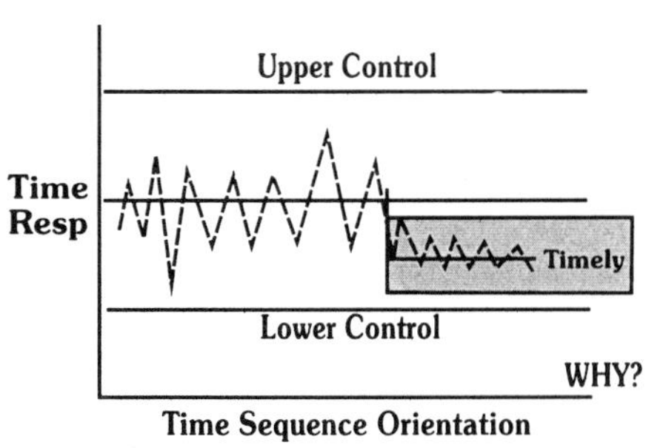

Time Sequence Orientation

Alarm System
AVE
$UCL = AVE + 3\ STD_{AVE}$
$LCL = AVE - 3\ STD_{AVE}$

When we are in control, we are running the machine in a steady fashion. In control is another way of saying that the machine is running with no change from the machine's typical running results.

Control charts, like the one to the left, show when a change has occurred for a particular machine.

WHY TIMELY REPORTING IS ESSENTIAL

In the early 70's, the comedian Flip Wilson had several characters that he portrayed in sketches. One of my favorite characters was Geraldine. She was well known for her lines, "What you see is what you get," in reference to herself and, "The Devil made me do it," when she was in trouble. She attended the Reverend Leroy's church that was named *The Church of What's Happening Now.*

Control charts are in many ways like Geraldine because the most important data point for a control chart is the most recent event or, as Geraldine would say, "What's Happening Now!" Conditions and changes in the past cannot be resolved and fixed but we can act upon the current situation.

Control chart alarms must be reported in a timely manner to allow us the opportunity to determine the reason why the change has occurred. Control charts demand that all results be posted immediately.

TACTICAL SUPPORT FOR THE OPERATOR

In a manufacturing setting, the operator is our tactician. The operator must use the available means to its optimum in order to run the facility as correctly and consistently as possible. Control charts alert the tactician to change, and are required to support the tactician's mission of consistently running the available facility. These statistical tools are based on averages and the use of standard error.

WHAT MAKES A CONTROL CHART

A control chart is the alarm system to note changes in a product or process function. The control chart monitors the various quantitative aspects of the function. This is a pictorial change-alarm system.

The control chart has an alarm mechanism calculated according to the actual performance of the operator's machine. The three lines that you see on a control chart are the centerline, and the upper and lower control limit lines.

The chart below shows the three lines of the control chart. The actual results are then posted on the chart to detect a change in the machine's function or performance.

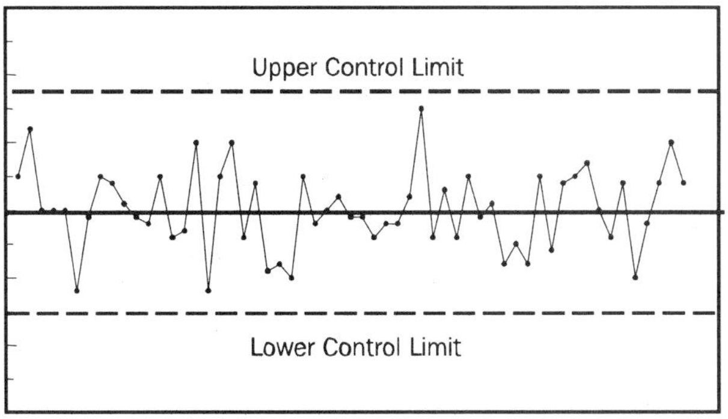

A control limit is used for the specific purpose of assisting the operator run a machine consistently. A control limit was defined in the 1920s. The original developers of SPC defined the term control limits to mean a specific idea. The intent was for the control limits to show the outer most spots where data is still allowed to appear. No data should fall outside of the control limits. Since control limits were defined so many years ago, any deviation from the definition is absolutely forbidden with no exceptions. This would be like trying to redefine the color red. We might be able to convince a few people to call red by the name of blue, but anyone to whom we had not talked would be extremely confused and rightly so.

The control limits should not be used to judge whether quality is acceptable. The control limits pertain to the correct and consistent operation of the machine. Data outside of the limits means there has been a change. A value outside of the control limits is not a call for action; it is a call for an investigation as to the reason for the change.

In addition, ALL interpretation rules are based on control limits that are computed based on the performance of the machine. No matter what the reason, we have too much invested in our proper control limit definition to allow any changes or deviations in the machine's performance.

MANAGEMENT'S ROLE

The concept of making what the customer wants is very important; we must be customer driven. At the same time, we can only evaluate capability when we are using the correct tool. Management must stay focused on strategic needs.

Management must understand its responsibility for providing a facility that will make what the customer wants. Management must understand that it will be held accountable for this mission regarding the capability of the machine. We provide very targeted tools to support operations' mission. Likewise, tools must be provided to assess management's mission of providing facilities that can meet the customer's needs.

When we say that we are capable, we are only talking about management and customer issues. The term capable means that the facility is making what the customer wants. There is a difference between the correct and consistent running of the facility and how well we are meeting our customer's expectations. When we understand the concept of capability, we realize that the objective of management is to assure that we are providing a facility that can make what the customer wants.

As management makes strategic decisions to plan for a goal, the limiting factor of availability that was placed on tactical decisions is removed. Management must find the means or facilities to supply operations. Strategic decisions involve determining the time, place, and resources required to accomplish our goal just as a general must obtain the resources, determine the place for the battle, and pick the time of the battle.

Strategic decisions are categorized as defining and obtaining the appropriate resources and then determining when and where to use them. Management uses strategies and strategic decisions to continually evaluate the adequacy of the facility for making what the customer wants. This strategic decision is based on the tool we mentioned previously, the capability study. If the facility is not capable, a strategic decision must be made concerning a proper course of action.

TOOLS TO SUPPORT THE MANAGEMENT/SALES/CUSTOMER ISSUES OF A CONTINUOUS OPERATION

There are many strategic analysis tools to support management/sales/customer decisions. One of these, a capability study, assesses how well what we are producing matches what the customer wants. A capability study would evaluate how well the product meets our customer's expectations. These studies are for a continuous process and assume the process is in control.

This chart is an example of a capability chart. Management uses a graph like this for reports:

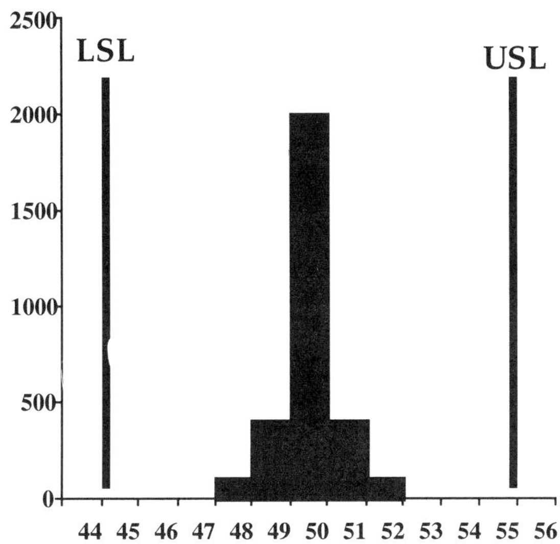

Because strategic decisions involve defining and obtaining appropriate resources and then assesssing when and where to use those resources, facility determination is clearly a management/sales/customer function. The capability study gives a strategist the information he needs to make appropriate decisions.

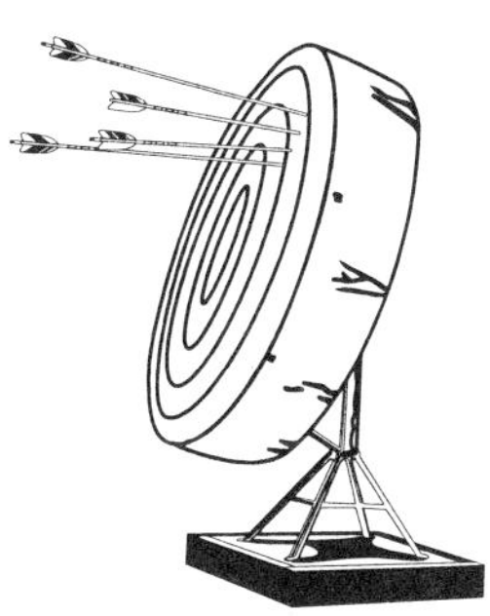

The archer assesses how the arrows are hitting the target and how many of them have hit the bull's-eye. Improvement is required to become a master. In control does not mean capable and capable does not mean in control. In control and capable are two independent functions. When we manipulate them with one another, spectacular things will happen! Later we will see how they form a team.

STRATEGIC SUPPORT FOR MANAGEMENT

Management decisions are strategic and assess impact. Management, or strategy, selects, assesses, and provides the means. All strategic analysis must support this mission. Since our management must assess the impact to the customer, customer specifications and the individual values must be used.

In a manufacturing quality setting, strategic issues deal with assuring that we have the correct facility to make what the customer wants. Thus we must compare our data to the customer's specification. To accomplish the strategic mission, the individual values must be used. To make the objective a more general statement, the strategist is responsible for providing the correct facility and also assuring that any change to the facility is capable.

THE STATISTICAL PROCESS CONTROL TEAM

SPC is comprised of the team that includes both control charts (for running the facility) and capability studies (assessing the impact). When working together and in harmony, the team is an unbeatable combination.

All changes are quickly detected by operations and then immediately reported to management. While operations investigate to find the cause of the change, management assesses the impact of the change on the customer. When both the cause and the impact are determined, then and only then can an action plan be plotted.

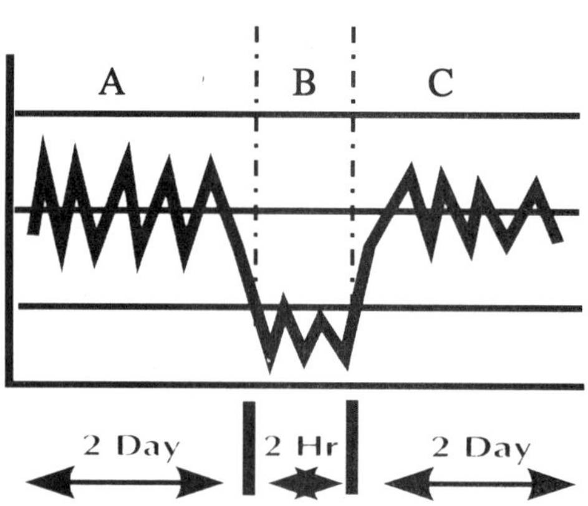

A + B + C = 1%
A = 0%
B = 6%
C = 0%

The chart to the left shows candy weights at the time of production. The weight of the candy has shifted from time period A to time period B. A second shift is shown from time period B to time period C.

The control chart keeps us from repeating the time-series grouping mistake that so often is made in capability studies. Only data that is produced during a time of statistical control can be grouped together in a capability study. In other words, we would not combine data from time frame A, with data from time frame B.

When we analyze the candy weights being produced for time period A, this capability study results:

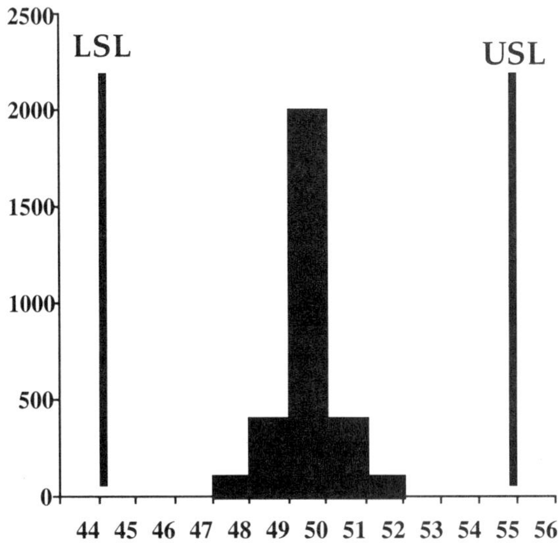

The histogram above is the results from time frame A with the specification limits overlaid. This gives us a look at how the candy production for time frame A would meet the customer's specifications.

Time frame A = 0% defective

The defective percentage for time period A is zero, for time period B is six, and for time period C is zero. Two distinctly different groups can be seen. If we attempted to group all the time periods together (A, B, and C), we would compute a defect rate of one percent which would be completely misleading and erroneous.

The tool to analyze this situation is called Statistical Process Control (SPC). SPC is the analysis tool of choice for a continuous operation.

The SPC team detects all changes and allows us to keep the improvements while getting rid of the deterioration. All changes, no matter how small, are addressed with equal fervor. Therefore, a series of very, very minute shifts are not allowed to slowly accumulate into one large problem.

Using capability studies and control charts, we can consistently hit the bull's-eye.

CHAPTER SUMMARY

Statistical Process Control (SPC) is the set of tools to monitor and control a continuous process. Control charts are the tactical tools to assist operations in the correct and consistent running of the facility. Capability studies are the strategic companion of control charts. They assess the impact of all changes on the customer so that management is never caught off guard.

The combined tools of control charts and capability studies allow our efforts to be focused to improve the product and the process. The SPC team can now support both strategic and tactical decisions with very specialized decision-making tools.

Control Charts

The control chart is the operator's friend in accomplishing the tactical role of correctly and consistently running the machine. The following checklist summarizes the explanation of a control chart and provides guidance as to which situations require this tool.

Consistency Alarm System Checklist: Functions of the Control Chart

- Operator's Tool
- One Particular Machine
- Continuous Process
- Time Production Oriented
- Alarm zones
    - Average
    - Upper Control Limit -UCL
    - Lower Control Limit -LCL
- COWHN (*Church of What's Happening Now*)

Remember that control chart methodology is a charter member of the *Church of What's Happening Now,* commonly referred to as COWHN!

Control Consistency Alarm Chart System
Operator's Tool
1 Particular Entity

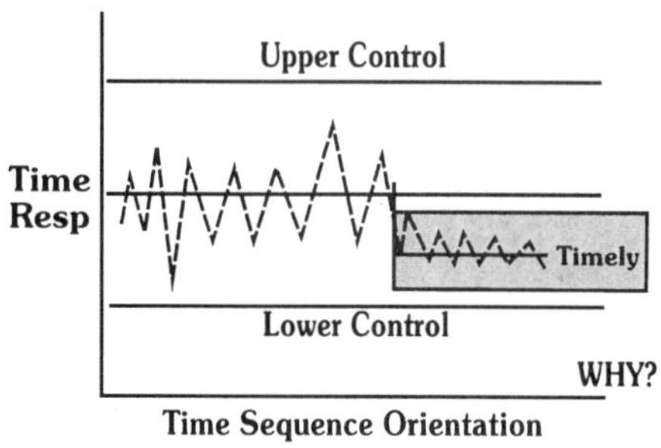

Alarm System
AVE
$UCL = AVE + 3\ STD_{AVE}$
$LCL = AVE - 3\ STD_{AVE}$

The chart to the left is an example of a control chart that has detected a shift. Operations will investigate the shift/change. Remember, a customer's requirements or specifications have no place on a control chart. This shift does not imply either an improvement or deterioration, but simply that a change has occurred.

Capability Studies

The capability study is management's tool to assess how well the facility meets the customer's requirements. The capability study must work with the control chart to assess all changes and to determine the impact of the change on the customer.

Control charts and capability studies form the dynamic team of SPC that is on a continuous search for improvements that are to be retained and deterioration that is to be eliminated.

Candy Cluster

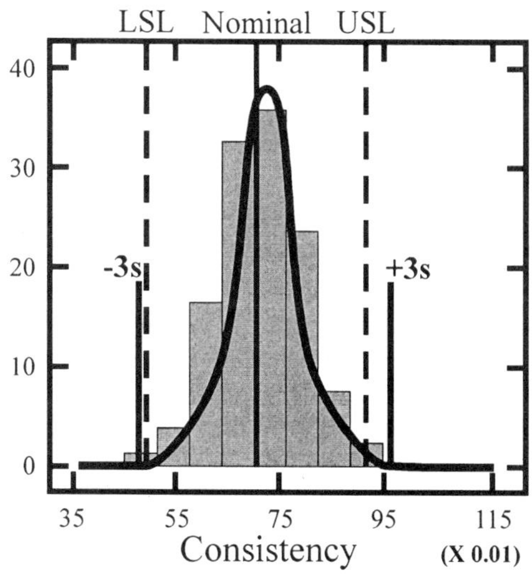

The chart to the left is an example of a capability study that is used to assess the impact on the customer. The control chart has detected a change and now our strategic decision-makers must assess the impact on the customer through the use of a capability study. Customer requirements and specifications belong on this tool.

This chapter provided a general look at control charts and capability studies. Since this book is targeted to operators, this book will focus on the operator tactical tool **control charts.** I will explain what is essential for an operator to understand and do to make a control chart a helpful tool for running his machine.

Strategic decision makers and implementers of an SPC program should read ***SPC: Practical Understanding of Capability by Implementing Statistical Process Control, third edition.*** This book explains both control charts (tactical decision-making) and capability studies (strategic decision-making) in considerable detail.

Chapter Two:
Measurement

WHAT ARE METRICS AND MEASUREMENTS?

Metrics are study areas about a given topic. Metrics are our means of capturing knowledge. We will have many metrics for which we must set targets, measure, monitor, and analyze. To deal with all of these study areas (metrics) we will need measurements.

Every time that we observe something we have the potential to learn. For most of us as we ride down the road in our cars we see but we do not retain. Think about your ride to work this morning. What exactly did you pass? Can you count how many signs were on the side of the road and what was on the signs? Most of us require a methodology to assure that our observations are truly retained and converted into knowledge. Measurement is this method.

Measuring our metric is key to learning. Back in the 1800s Lord Kelvin said that if you can measure what you are speaking about, and express it in numbers, you know something about it. But if you cannot measure it and you cannot express it in numbers, your knowledge is meager and unsatisfactory.

Measurement adds to and builds upon the knowledge we get by simple observation, by just looking. When we measure, we observe, make a comparison, and make a written record. All measurement is the systematic comparison of observation to a standard. All *effective* measurements are recorded.

Measurement can be automated, and observation sometimes requires special tools to help us observe things below the threshold of direct human perception.

The **metric** is what we are studying and **measurement** is how we study.

HOW DO WE MEASURE?

My daughter is looking for her first house. In her first look at houses she simply saw what she felt was visually appealing and said, "I want to purchase it." Then questions began to arise like will her current bedroom suite fit in the master bedroom? Is the kitchen big enough? Will the color of the living room walls match her furniture and paintings? Is the living room big enough for her furniture? She spent hours seeing the house. She knew that she liked the house but as Lord Kelvin said, her knowledge of the house was meager. The solution to my daughter's problem was measurement.

As my daughter began to understand the complexity of buying a house, she began to focus on one problem at a time. Narrowing her focus to the living room, her concerns began to revolve around the size of the room.

The first effort at analyzing the room size was simply to pace off the living room. My daughter's boyfriend stepped off the room, which was five paces wide and nine paces long.

We began talking about how she was going to use the room. Where would she put the couch? Where would the television go? Would the room need to be repainted? After fifteen minutes the discussion came back to the size of the room. Did anyone remember the size? I said it was eight steps by seven steps. Her boyfriend chimed in because he was certain the dimensions were six paces by nine paces. My wife and daughter both thought the dimensions were something else entirely. This led us back to having to re-step the room. One vital point of measurement had been omitted – a written record.

To assure that this time we got a good reading, I said that I would do the walking. Remember, the first steps were by my daughter's boyfriend (height 6' 4) and these readings were by me (height 6'). As I stepped my first few paces, I realized that I was shorter than her boyfriend. So I started over and began to compensate taking huge giant steps. After three attempts the family agreed that a better method must be found.

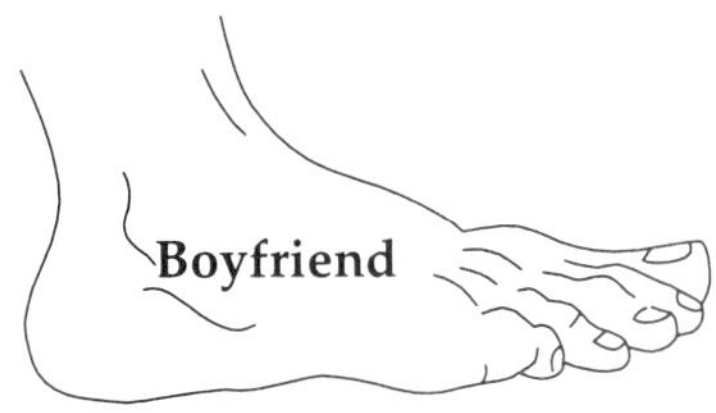

My daughter suggested that we use a heel to toe count of the number of feet across the room. We measured with my daughter's boyfriend's foot. Remember that his height makes his steps giant compared to mine. Since he is so much taller than I am, his feet are also much larger.

We now religiously write down the living room dimension readings of 21 feet by 11 feet. Actually, the true measurement was 21 boyfriend feet by 11 boyfriend feet.

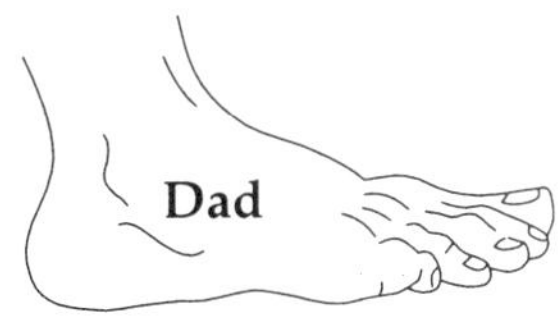

The boyfriend drove back to his house, and we drove back to our house. Once home, we decided to compare the possible living room against our living room of 23 years. With no boyfriend available, I said "no problem" and measured the living room with my own feet. The measurement of our living room was 30 Dad feet by 25 Dad feet. Now we began to discuss how our existing home matched the potential house.

Next my daughter wanted to compare furniture that was in storage. She went to the storage shed to get measurements of her sofa, two Queen Anne chairs, coffee table, end tables, and television cabinet. All of these measurements were made in daughter feet.

We had three sets of feet that we were using to make our measurements. This made any accurate comparison impossible. What was needed was a standard foot or, better yet, a standard measure for lengths.

STANDARDS

From this experience you can now easily see the need for a standard. A standard for length can be the English or Metric systems. English uses inches, feet, yards, and miles, while metric uses millimeters, centimeters, meters, and kilometer. Each unit and system has a very clearly defined standard. No matter where or who is doing the measurement, an inch is always the same length.

Now when we go to measure my daughter's potential house we have selected English feet as our standard unit of measurement.

Standards are required for every type of measurement. Weights, density, money, moisture, etc. will all require the dimension to have a standard. Weights have standards like pounds or grams. A pound is a clearly defined standard. All measurement units (inches, feet, meters, pounds, etc.) have a standard. In the United States, the National Bureau of Standards, a federal agency, maintains these standards.

Gauges

Now with the standard for measurement selected, a gauge must also be developed. The gauges will have names like ruler, micrometer, etc. The manufacturer calibrates every gauge when it is produced. Some types of gauges will need to have their calibration re-certified. The need for re-certification depends on the accuracy requirement. For the purposes of measuring the house we will use a ruler like the one below.

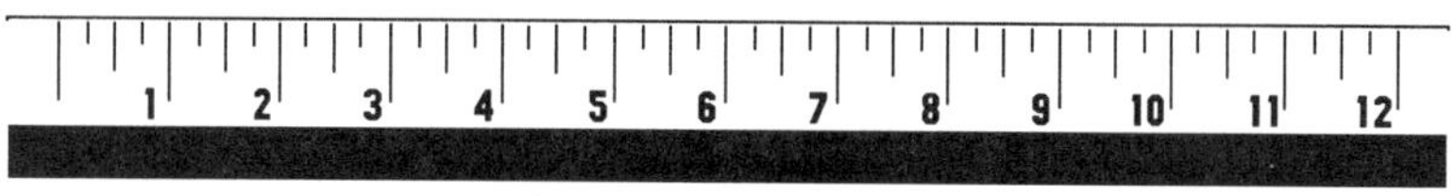

A weight metric requires a different gauge. A scale is this gauge and weighs the mass in units. The scale must be calibrated to give the readings of a standard like pounds.

Crude, simple, and precision measurement gauges

The measurement device that we use may take many different forms. The weight of an object can be made with a simple balance, floor scales, or a precise analytical scale. The level of precision required is dependent on information the gauge is providing. For my body weight to track my diet, a simple bathroom scale would work well, but the amount of nuclear fuel in a nuclear power plant requires much more precision.

Crude measurements are not to be confused with bad or inaccurate measurement. Some metrics do not require extreme accuracy. When my daughter does buy a house, she will need to know some general lengths of the lot and the house exterior to buy fertilizer. A crude technique of pacing off the lengths will work fine for her general purposes.

To measure the interior dimensions of the future house, more precise readings are required. A simple measurement could be a tape measure to measure the interior walls, doorways, and windows.

This crude measurement would not work for purposes of getting the loan on the house. The deed will require more precision and the introduction of a surveyor and his precision tool, the transit. The transit is a telescope with various attachments for measuring lengths and angles.

Machined parts requiring precision to the thousandth of an inch requires a micrometer. The micrometer is a device used to measure length or thickness. If you have worked with hand tools, the micrometer will remind you of a "C" – clamp. However, it is a delicate, precision instrument and it should be handled carefully.

Micrometers are manufactured in different sizes and styles. The most common are specifically intended to measure dimensions from 0.0000 – 1.0000 inches. Newer models have digital displays. A digital display does not fundamentally change the accuracy or precision of the instrument. It simply makes it easier to observe a reading.

To summarize, measurement requires a method for how we measure, a standard unit, and a calibrated gauge. Each reading must be recorded. Each metric will dictate what level of precision is required. Too much precision can be as detrimental as too little precision.

WHAT TO MEASURE?

In my daughter's house hunting experience, her reason for measuring was to see if her furniture would fit in this new lovely house that she wanted to buy. She had worked hard to save her money so she wanted to make sure she made a wise investment. The measurements that you will be taking are so that you can run your operation or machine to the best of your ability.

Knowing what to measure is vital to running an operation, making improvements, or understanding what is going on. The First Principle of Process Management is the key to knowing what to measure.

In my book, *Optimize Your Operation*, I told the story of my mother learning the art of biscuit making from my grandmother. Since Grandma was an accomplished biscuit chef but my mother was not, the knowledge transfer was very difficult. Without the proper analytical tools used in the correct areas, Mama burned many biscuits and the baking knowledge was not transferred. Mama was finally able to master the art of biscuit making when she learned and used the Principles of Process Management. Here is a brief summary of the First Principle of Process Management.

FIRST PRINCIPLE OF PROCESS MANAGEMENT

A fundamental understanding of BOTH the product and process is essential to improvement. Both the product and the process must be described and understood individually and separately. The underlying component for improving the product is the process.

When Mama measured the correct metrics, the knowledge of how to bake was learned. You must make sure you are running your machine as efficiently as possible, and measurements are the way to do this.

Don't forget, the First Principle of Process Management is crucial to improvement. There are two broad areas: product and process. We must have data from each area, and in order to improve the product, we must understand the process better.

A product is...
- Noun
- Result

When we are discussing products, we are focusing on the result of our activities. The product is always a noun.

A process is...
- Verb
- Activity
- Something that builds the product

Process discussions are focused on the activities that build the product. The process is always a verb.

The Walkabout® Dependency Diagram is the tool designed to help us understand the product, the process, and their relationship to each other. Walkabout® construction is covered in detail in *Optimize Your Operation*.

Since the product is the goal and objective, many people monitor only the product, but to achieve the goal, an equal knowledge of the process is required. A clear understanding of the First Principle adds focus to what we monitor. The importance of product and process knowledge must be clear because metrics and quantitative measures are vital to understanding the product and process. We will learn how to identify and measure metrics from the tool called the Walkabout® Dependency Diagram.

The diagram provides the structure so that we will never skip a step or forget a component. The first step is to define and document the product. The second step is to define and document each step in the process. We write this information on the front of the Walkabout®.

With this information, we have started on our journey to improvement. Many people have difficulty making a clear distinction between product and process. The important issue is to monitor both with equal fervor. Over-emphasis on either one will result in mistakes.

THE IMPORTANCE OF PROCESS DATA

When we monitor the product only, we are aware only of its deterioration or improvement. However, this awareness of change does not give us a reason for the change. Thus product monitoring is important, but it is reactionary. When a product problem is encountered, we take action to correct it. The drawback to acting on a problem is that we must have a problem in order to act on it! Monitoring the product and taking action after a problem has been detected tolerates waste with no understanding of why it occurred. When improvements in the product are found, no knowledge of the cause is determined and the improvement cannot be repeated.

WITHOUT KNOWLEDGE OF "WHY", WE CANNOT IMPROVE

For us to make Grandma's biscuits, we must understand why things occur the way they do. Product information gives no insight into why we made good biscuits or why we burnt the biscuits. Since we don't know what causes good biscuits, we cannot repeat the activities that produced them. *This loss of knowledge is tragic.*

When we monitor the process and the product, the process parameters allow us to know the conditions that caused the change. Monitoring both product and process is essential for proactive decision-making. With product and process knowledge, measurable improvements and elimination of waste will occur. Understanding the cause of change will increase knowledge, leading to improvement.

We need to keep our focus on the process if we want to improve the product. Too often we use SPC to monitor the product. The "P" in SPC stands for *Process*, NOT *Product*. Remember to use SPC (Statistical **Process** Control) to monitor both the product **and** the process metrics.

THE KNOWLEDGE OF "WHY" ALLOWS US TO IMPROVE CONTINUOUSLY

By adding knowledge about the activities required to make a biscuit, we begin to learn what is required to make good biscuits. The information about the different processes that were used for good biscuits versus defective biscuits allows us to continue to repeat the improvements and to discontinue the defects.

In the perfect world of my grandmother and her biscuits, I never saw her check the biscuits when she removed them from the oven. She kept her focus on the process. Since she knew all the subtleties of baking, she took corrective action as the biscuits were baking in the oven, so that, in a proactive manner, she avoided ever producing any burnt biscuits. She was a perfect chef; unfortunately, most of us are not perfect chefs of our processes.

Often we spend a great deal of effort monitoring defects. When we are talking about defects, we are actually discussing the product. The problem with focusing on these undesirable product characteristics is that once the information arrives, it is history because the product has already been produced. If a defect is present, we now have waste, unhappy customers, scrap, and re-work.

Since we live in a not-so-perfect world, and our knowledge of our process is not nearly so complete as my grandmother's was about her oven, we must continue to monitor both our biscuits and our oven with equal intensity. More precisely, we must monitor both process and product.

As we gather data and knowledge, we must accumulate equal amounts of knowledge about the product *and* the process. This will allow us to build the knowledge base for both.

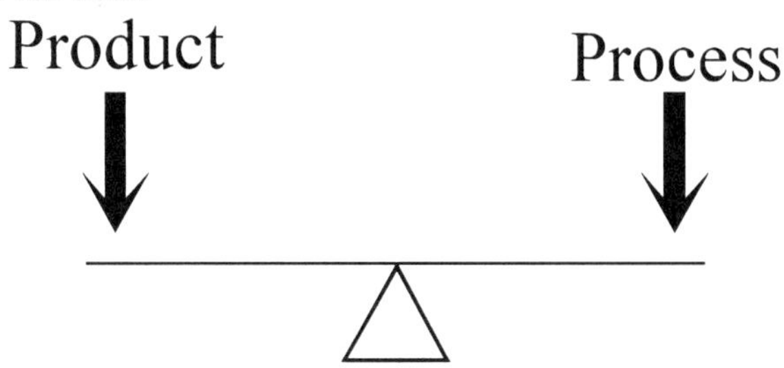

The First Principle explains the interconnection between product and process. We cannot improve the end product without improving the process, and in order to improve the process, our knowledge base must be dramatically increased. In this chapter, we will draw a clear picture of the product/process relationship as a system of interacting and intertwined components. To simplify our efforts, we will use a simple technique to draw a schematic of the system. The system picture is called a dependency diagram and it is part of the Walkabout® we discussed earlier.

HOW DOES THE WALKABOUT® DEPENDENCY DIAGRAM LOOK?

The chart below is the Walkabout® Dependency Diagram for baking Grandma's biscuits. The dependency diagram shows the sequence of activities. Let's use this example to explain how to read the Walkabout®.

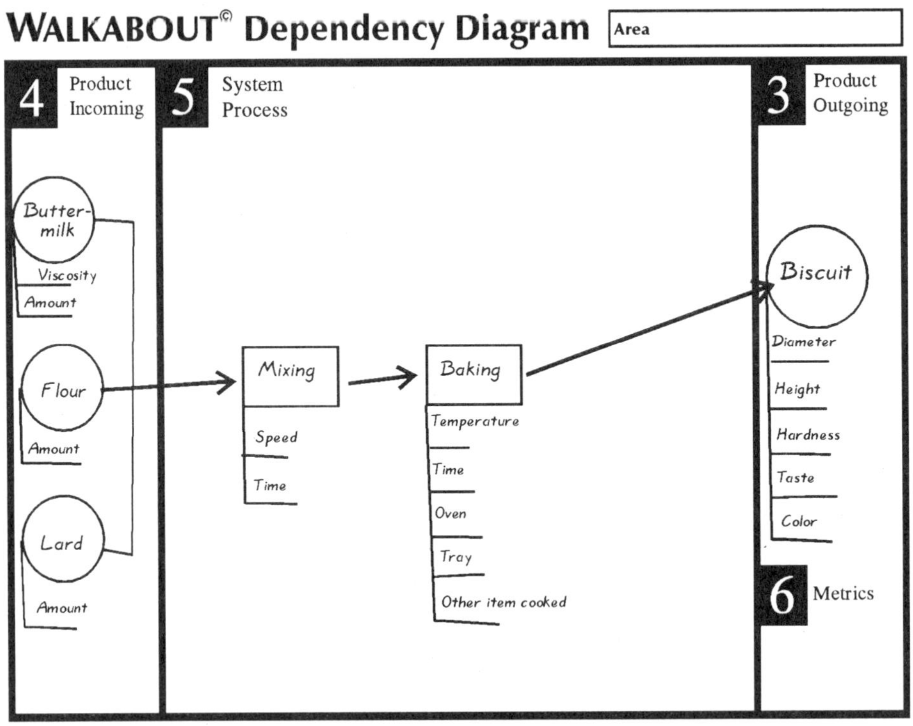

The dependency diagram is shown with circles for the products (think of biscuits) and rectangles for the processes (imagine an oven for baking). The schematic shows the sequence in which each activity occurs. For example, the diagram clearly illustrates that we mix before we bake. The dependency diagram makes the learning process easy and easy-to-follow.

METRICS ABOUT THE PRODUCT AND PROCESS

Metrics are study areas about the product and the process. Metrics are our means of capturing this product and process knowledge. We will have many metrics for which we must set targets, measure, monitor, and analyze.

Hanging from the products and processes are legs where we list the metrics that we will want to study. The dependency diagram and the metric legs form the blueprint.

The Walkabout® provides a communication medium for each and every associate to learn, share findings from bottom to top, and retain this knowledge for current and future associates. This communication tool will allow us to:

- **Learn** our product and process.
- **Share** our knowledge and what we have learned with other associates.
- **Teach** and train others in what we have learned.
- **Have a checklist** to assure we don't forget any of the details of our knowledge.

The Walkabout® gives us one common language and framework to share and increase the knowledge that we will learn. Failing to have a common language leaves us in a chaotic state, but a common language gives us a method to get the best knowledge from all associates. Progress and knowledge will not occur if we can't communicate using a common language.

The Walkabout® describes dependencies and defines metrics. With the Walkabout® in place, we have a tool that gives us a short and long-range plan for metrics, which will guide us in all our studies. Quantitative analysis is required for a vivid picture of the product and process. Techniques required for defining metrics, measurements, and targets are explained in this section.

USING THE WALKABOUT® AS A METRIC BLUEPRINT

The Walkabout® Dependency Diagram is the comprehensive tool to evaluate how well the effort has been described and also becomes the tool to monitor compliance.

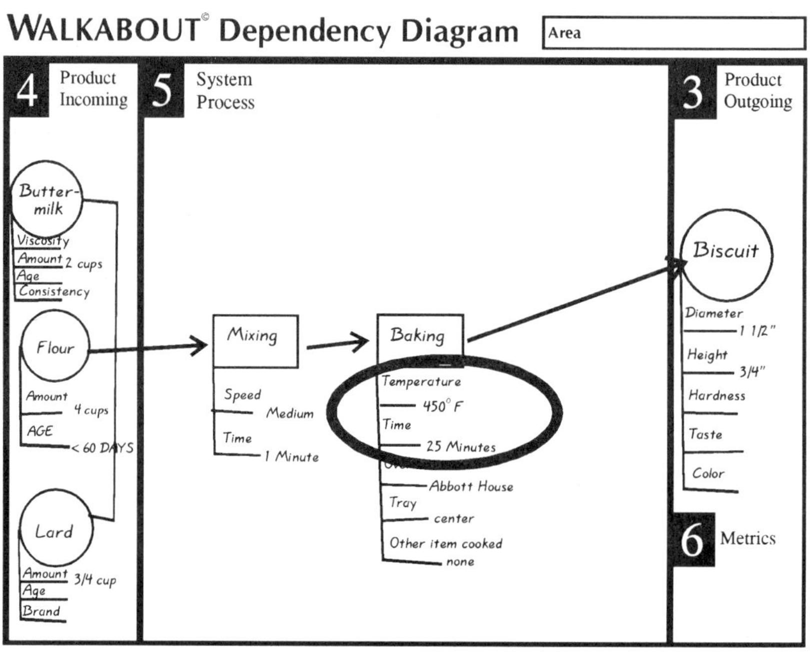

The Walkabout® is a tool that allows us to make use of the first Principle of Process Management in order to document and improve our particular product and process. Again, the dependency diagram documents the product and process and is driven by the first Principle of Process Management.

The Walkabout® Dependency Diagram is the tool to examine the system of incoming products, manufacturing processes, and outgoing products. It shows the sequence of processes and their dependencies. Metrics (study areas) are shown in relation to the products and processes.

Being able to see all the metrics in picture form provides a mechanism for identifying which metric to monitor. The Walkabout®, as a metric blueprint, provides the picture for selecting which metrics to start monitoring, which future metrics to monitor, and which measurements to determine. This allows us to understand and monitor our progress of current and future information needs.

My Grandmother used the Walkabout® to clearly document the products, the process, and the product/process relationship for baking biscuits. The Walkabout® also contained all present and future product and process metrics, measurements, and target values and was first and foremost a tool to document how Grandma made the biscuits. Grandma used the completed diagram to teach my mother how to make southern biscuits.

When a metric is selected, we have a place to start our efforts and are ready to begin Statistical Process Control.

As Mama and Grandma worked to improve the biscuits, the Walkabout® grew and grew. Both the process and product understanding are shown on the form. This tool provides the medium to share the knowledge with future generations so that the legacy of southern biscuits will live on.

We must strive to accomplish the levels of expertise in our process that my grandmother and mother achieved in their biscuit-making production. Grandma and Mama did not achieve their expertise by easy and lazy efforts. Improving our knowledge base is a worthwhile objective, but it requires hard work on the appropriate things. It is not enough to know what to do. We must actually do it. We must build and understand the tools of process management, use our knowledge, and execute our plans.

CHOOSING WHEN TO USE SPC

A continuous operation is one in which the same activities happen over and over again. Many operations fit these criteria. Statistical Process Control (SPC) is a set of tools that was designed specifically for this kind of situation. First we must determine which metrics are continuous and thus a candidate for SPC. Once the type of situation has been determined, we then select the appropriate statistical tool.

WHAT DID MAMA CHOOSE AS METRICS FOR SPC?

Mama knew that SPC was a tool to assist in monitoring a continuous operation. According to the completed Walkabout®, some of Grandma's metrics are continuous in nature. She will need to monitor these metrics with SPC, our tool of choice for continuous operations. The completed Walkabout® guides us by keeping all of the metrics organized, so that we can pick our first monitor area.

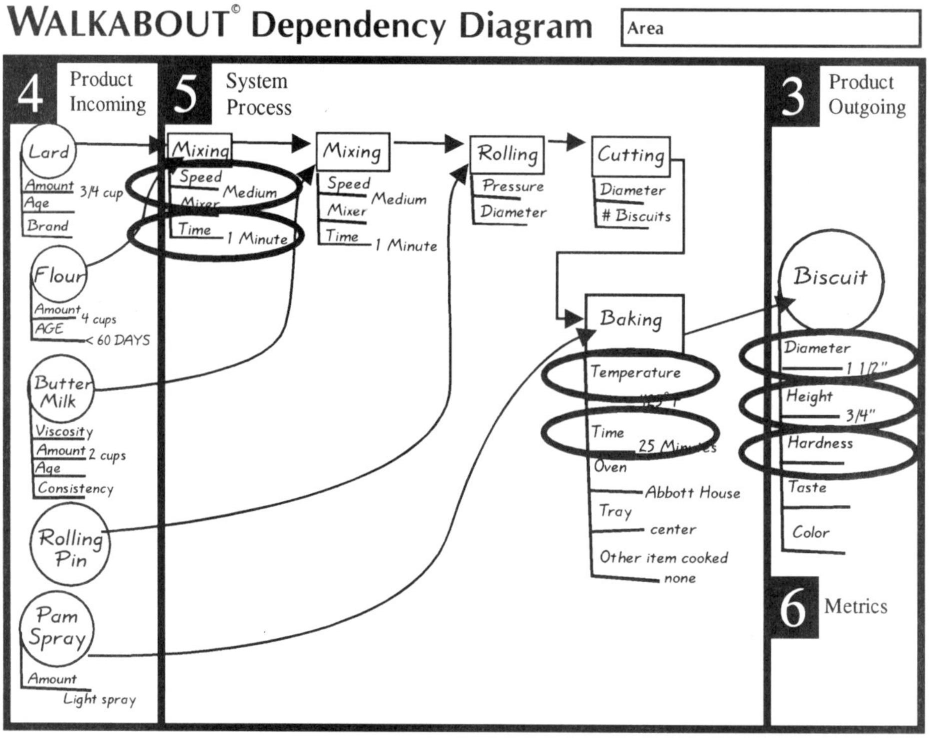

Oven Temperature – Grandma and Mama believed that oven metrics are important to good southern biscuits. A continuous surveillance method must be used to monitor the oven temperature so that any fluctuations are noted. SPC is the tool for the task of continuously monitoring the oven temperature.

Other biscuit and oven metrics may be candidates for SPC monitoring — these metrics include the mixer speed and length of mixing time; the oven duration; and the biscuit diameter, height, and hardness. We would have control charts to assess the consistency of each metric. Each metric could then be compared against the specifications and targets to determine the capability of the process. Once a metric is selected, we must learn to measure the metric.

HOW WILL WE MEASURE THE METRIC?

Measuring the metric is essential to the learning process. If we never shoot the arrow, we will never know for certain how good our archery skills are. Similarly, taking readings is essential to our learning efforts. No measurements mean no knowledge.

Measuring our metric is key to learning. Remember that Lord Kelvin said, "When you can measure what you are speaking about, and express it in numbers, you know something about it. But when you cannot measure it, when you cannot express it in numbers, your knowledge is of a meager and unsatisfactory kind."

When an archer shoots an arrow into a target, the different locations of the arrows in the target paint a picture about how he should improve. When the target includes numbered circles, the knowledge of the archer's accuracy is increased. Adding numbers to the target advances our ability to learn. We can measure how many arrows fell in each area to obtain a better picture of our archery skills. Measuring against a target allows us to understand both how consistent we are and the capability that we possess.

This consistency and capability knowledge becomes essential for any business or operation. To achieve the consistency and capability knowledge, measurements are required.

WALKABOUT® AS A MEASUREMENT BLUEPRINT

The completed Walkabout® will provide the blueprint for all measurement definitions.

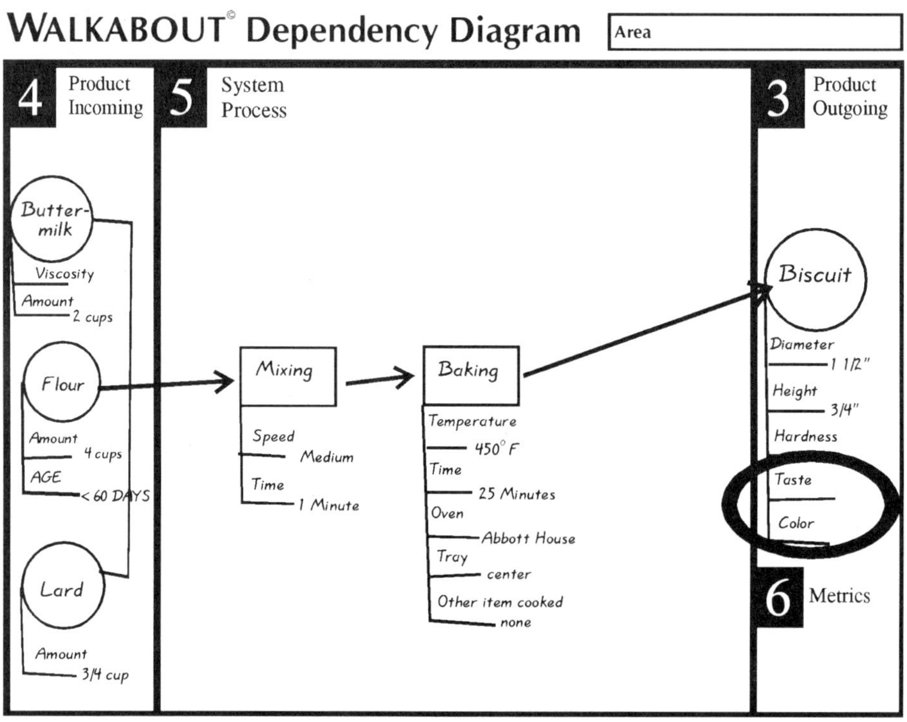

Some metrics will be either difficult or impossible to measure while other measurements may also be very expensive. In these situations, we do not want to lose our metric, but we may simply defer our measurement until taking it is feasible. With a blueprint of our metrics and measurements, we will never lose sight of our need to gather more information, and we will also know what information is missing. In the example above, technology is not available to measure taste or color. These are still metrics, but they will be added when the technology is available in the future.

WHAT MEASUREMENT TYPES ARE AVAILABLE?

There are two metric measurement types. The first type, called **attribute**, has counts, pass/fail, or true/false as its properties. Attribute data is also referred to as categorical data. The second metric type has measurable properties and is called a **variable**. Variable data is also referred to as continuous data.

ATTRIBUTE DATA

Attribute or categorical data is pass/fail and counts. This type of data is often used when deciding whether or not to accept a part in manufacturing, and for some service industry (administrative) tasks. Even though this type of data is pass/fail, we can count the frequency of occurrence. Attribute data is "yes" or "no," "good" or "bad," "go" or "no go." Another example of attribute data is the count of the number of red M&Ms in an M&M bag.

How we measure the metric "flour age" can take on several forms. Attribute data could be one possible choice for determining the flour age by monitoring the number of bags produced before a certain date.

The most elementary type of attribute data determines whether an item is good or bad. We might count the number of biscuit batches in which the flour age exceeded a pre-specified number of days. If we set the target age to be less than 30 days, our measurement would then be the number of batches over 30 days old.

VARIABLE DATA

Variable data is a continuous measurement. Examples would include temperature, length, width, diameter, shading, and days. Variable data (also known as continuous data) can be recorded on a scale: meters, grams, inches, or viscosity.

Units are an example of variable data. Examples of this are length, width, time, and weight. We might want to know the flour age in number of days since it was milled.

We also might measure against multiple units. An example of this is density, which is weight per volume. We want to know the flour age but rather than simply monitoring number of days, we want to divide the number of days by the number of biscuits in the batch.

Several methods and types of measurement may be available to us. During our flour age discussion of data types, many methods were discussed but more are available. The objective of metrics is to help decide how to measure and what data type to use.

STEPS TO METRIC MEASUREMENTS

There will always be several metric measurement options available to the strategic decision-maker. Consideration must be given to the amount of information to be gleaned, the cost of the measurement, the accuracy of the measurement, and the difficulty of the test procedure to accomplish the measurement. Over time, I would expect the technique for gathering the measurement to change and improve. These improvements should not be confused with more costs or just more precision; instead the improvements must increase product and process knowledge.

Determining the kind of measurement (attribute or variable) and the technique for gathering the measurements are key to increasing our knowledge base. Our strategic decision-makers establish what metric measurement we must take and continually refine and add to these measurements.

In addition to the obvious things like temperature and duration, metric measurements must be established for our housekeeping cleanliness. For example, a frequency check for the cleanliness of the oven must be established because this is a vital metric measurement. Similarly, checking for incorrect operations must be a regular part of the housekeeping effort. It is reasonable then that these checks for incorrect operation must be included in the Walk-about®.

Strategic decision-makers first determine the measurement technique and procedure for each metric. The intended use of our information will drive our test method. The need for different measurement methods to support operations versus protection of the customer should come as no surprise.

Once the strategic decision-makers have determined the measurement and the procedure for taking the measurement, we, the tactical operation, must have a clear definition of the measurement's mission.

A clear measurement mission is essential to the successful gathering of product and process knowledge. The objective and purpose of the information is key to how to measure the metric. A measurement technique that requires eight hours to perform a test may be very acceptable for a quality assurance function but unacceptable for an operational function. Running a machine requires a much more timely response than a quality assurance function. Having multiple measurement techniques available, and having to decide which one(s) to use, poses a common dilemma; the context of the metric resolves this question.

Having a clear mission gives the metric purpose. With this purpose in hand, the measurement of the metric has a start and an end because when we have learned all that we need to know about the mission, the measurement chart may no longer be necessary.

MEASUREMENT ERROR

There are four types of measurement error: operator, gauge, method, and blunder. We must guard against each error type at all times.

Operator error is simply an operator mistake. An example is the operator not following the strategically defined method. The operator uses the wrong gauge. Another example is bad or poor operator training. The prime error, in this case, is coming from the person (operator) doing the measurement.

An example of **gauge error** is when an improperly calibrated gauge, like a scale, gives a reading five pounds too heavy.

If everyone does not use the same *method* of measurement then measurement will not agree. If two carpenters measuring a bookcase cannot agree on whether to take an inside or outside measurement, then their results differ.

Blunder is a gross error in method. When there is evidence that a blunder probably took place, the results should *always be thrown out!* This is not permission to throw out what is call "outlier" data (wildly varying results), unless there is probable evidence of a blunder in measurement method.

VARIATION IN MEASUREMENT

From operator to operator, with correctly executed methods and good gauges, one operator's reading may be slightly different from that of another operator.

From one gauge to another gauge, despite using correctly executed methods and correctly calibrated gauges, our readings may be slightly different.

As we correctly execute our method, the method may still have room for some level of inconsistency. When one operator uses the same micrometer, following the established correct method, a slight deviation may appear simply from the amount of pressure exerted on the gauge dial.

Random cause does *not* mean "without cause." Random cause is the total effect of an uncountable number of individually imperceptible, independent, unknown things. Random cause is usually not a problem with "rough carpentry" measurement. It is a problem of precision measurement, where the object of observation is below the threshold of direct human perception (i.e. too small, too big, too slow, too fast).

MEASUREMENT ERROR VERSUS VARIATION

Error is a mistaken reading. Variation is deviation from the correct value (if the correct value was known). An error occurs only by mistake while variation is the unintentional differences of readings.

Often the terms error and variation are used interchangeably. This creates confusion and mistakes! The big distinction between error and variation is, error can and must be avoided. Variation is the natural deviation of a measurement. As our process improves, the variation may be reduced.

VARIATION AND THE RANGE

The following are actual readings of a metal rod's diameter, using a micrometer.

0.0497	0.0508	0.0506	0.0502
0.0503	0.0509	0.0495	0.0505
0.0490	0.0496	0.0487	0.0497
0.0503	0.0492	0.0497	0.0499
0.0496	0.0510	0.0494	0.0495

In the sample data, you can see that when 20 people measured the same object with the same micrometer, the results varied. It is true, some people happened to agree. "0.0497" was reported more than once.

Notice, also, that there does not seem to be a pattern. **Random cause variation** is like that. If we knew the pattern, we would compensate for it, but high readings are sometimes followed by more high readings, sometimes by low readings, sometimes by middle readings. First, there is no pattern. Second, we really don't have a feel for what is "high," "low," or "middle" until we take multiple readings!

Notice the spread from smallest to largest reading. This is one simple indicator of the precision of this measurement. You would not want these folks responsible for sorting out the large and small of 0.0499 and 0.0501-rod diameters, unless they used better methods and tools. They could sort 0.51 from 0.54 inch parts with the same methods and tools used in the example above.

VARIATION AND THE RANGE – GRAPHIC

It is helpful to graph the "spreading" of results due to small variations in method, random cause variation, and blunder.

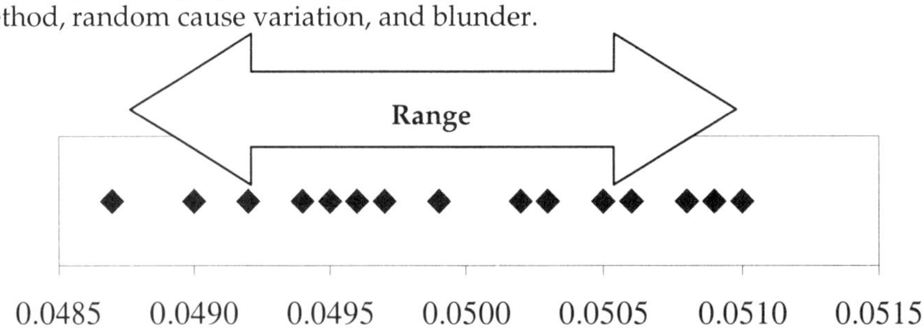

The pattern in our sample data is typical of repeated measurement. The spread (range) could be more if people were lax in method, and the range could be less if they were strict in method. There is no obvious indication of a blunder in the data, and the people who made the measurements reported no error.

If someone had reported, "The part I measured was dirty," or "The micrometer was hard to adjust —it was rusty and bent" that would certainly be evidence of probable blunder in measurement, and the readings should be re-taken with correct methods and tools in which we have confidence.

CENTRAL TENDENCY AND THE AVERAGE

The average of repeated measurements can be a better estimate than a single individual value. Sometimes a measuring process with inadequate precision can be "sharpened up" a bit with repeated measurements.

0.0497	0.0508	0.0506	0.0502
0.0503	0.0509	0.0495	0.0505
0.0490	0.0496	0.0487	0.0497
0.0503	0.0492	0.0497	0.0499
0.0496	0.0510	0.0494	0.0495

An **average** is simply the sum of all the values divided by the number of values. For our example, the sum is .9981 divided by the 20 data points. The average is then .0499.

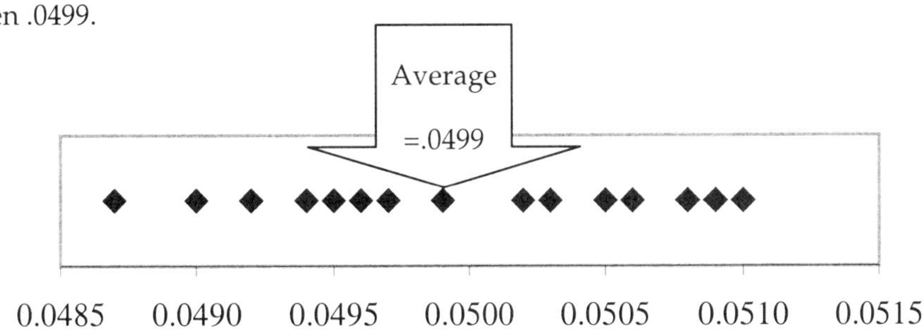

Clearly, the precision bought with multiple readings comes at a higher and higher price. Unless they are conducting a measurement study, most people do not bother with repeated measurement over more than five readings. If the precision is inadequate, they find a better way to measure.

The average is monitoring a value called central tendency. Central tendency tracks the center point of our data.

A PICTURE IS WORTH A THOUSAND NUMBERS

As we begin to read and interpret our numbers, simply looking at the number is important. The following set of numbers are readings of our biscuit hardness. There are 25 numbers. Simply reviewing the numbers gives us information, but all the numbers are very close.

	Data
1:00 PM	11
2:00 PM	10
3:00 PM	10.25
4:00 PM	10.5
5:00 PM	9.75
6:00 PM	9
7:00 PM	10
8:00 PM	11
9:00 PM	10.5
10:00 PM	10.25
11:00 PM	9.75
12:00 AM	10

	Data
1:00 AM	11
2:00 AM	10
3:00 AM	10.25
4:00 AM	10.5
5:00 AM	9.75
6:00 AM	9
7:00 AM	10
8:00 AM	11
9:00 AM	10.5
10:00 AM	10.25
11:00 AM	9.75
12:00 PM	10
1:00 AM	13

The table of numbers above clearly fits Lord Kelvin's criteria for measuring, since we are able to express, in numbers, what we are talking about. The only dilemma now is being able to distinguish the numbers. Simply reviewing the numbers, no clear difference can be made.

Now comes our new tool — pictures, or better yet, a chart. There are many kinds of charts such as line, bar, and pie. Since we normally are working in a time-oriented situation, a line chart will keep the sequence of events in chronological order.

To better detect a difference, a chart is very valuable. The chart below is one of many that can help us.

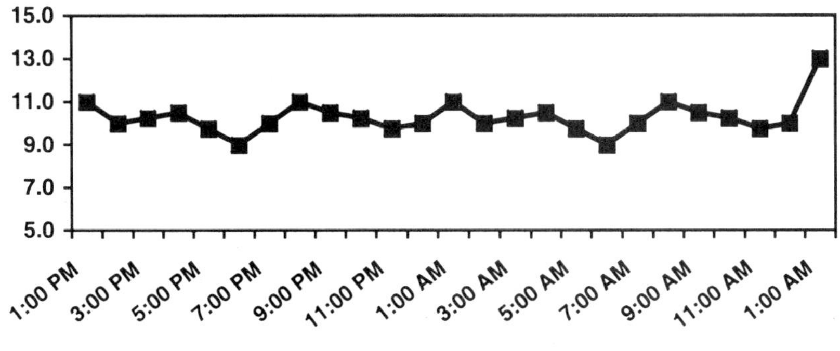

From the chart we can see that the typical reading-to-reading variation has changed in the last reading. Note how much easier the difference is to observe on the chart than by looking at the table of numbers.

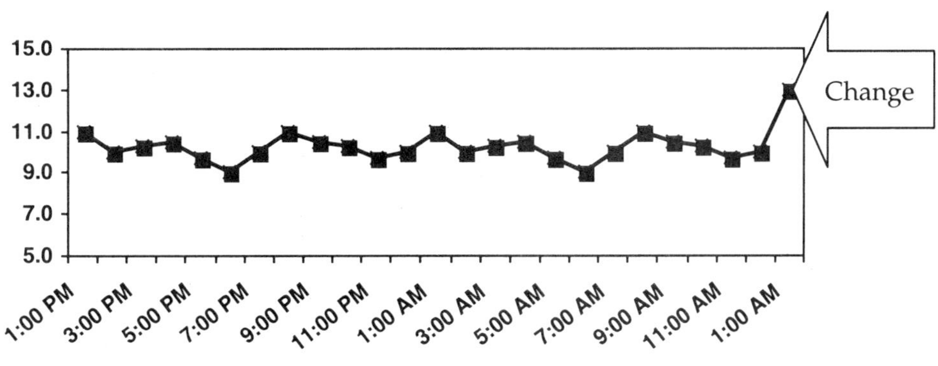

Adding features to our chart, a set of lines called control limits are drawn. Control limits are calculated values that represent the typical norm for the data we are plotting. With the control limits plotted, our job of detecting difference and change now becomes much easier.

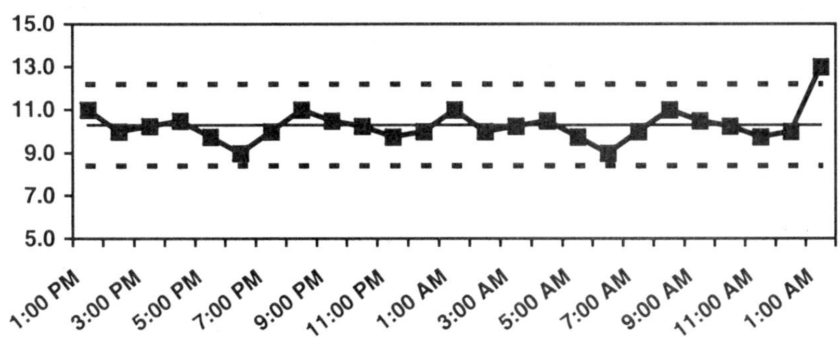

The last reading is clearly outside of the typical values as shown by the control limits. Thus the combination of chart and control limits forms a truly helpful tool. This tool is called a control chart. A control chart is a tactical tool to help an operator run his machine to maximum performance. Be aware that a control chart and its limits must be calculated for each individual machine. This is because each machine has its own normal running environment.

HOW WILL WE GATHER OUR MEASUREMENTS?

Since control charts are tactical tools, clusters of readings will be captured. SPC is a tool for a continuous process and, with this in mind, our readings must be captured in chronological sequence. The readings will be pulled to form a grouping called a subgroup. The subgroups must maintain their time sequence and have equal amounts of time passed between each of the subgroups.

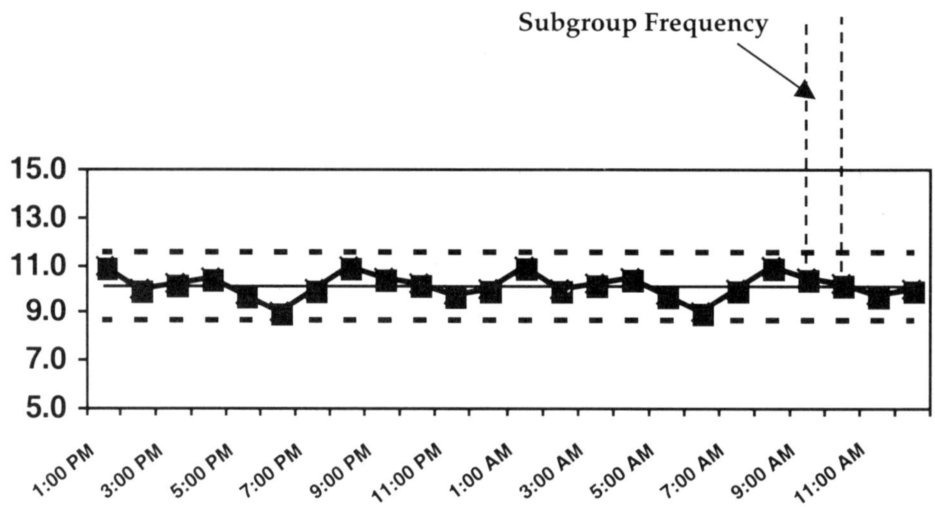

The spacing between each subgroup should be the same, if at all possible. The time spacing is called the subgroup frequency. Samples of product must not be drawn haphazardly and must not introduce bias toward either good or bad. As nearly as possible, a single subgroup must include product or process readings produced at the same time under the same conditions.

CHAPTER SUMMARY

Measuring our metric is key to learning. Lord Kelvin said, "When you can measure what you are speaking about, and express it in numbers, you know something about it. But when you cannot measure it, when you cannot express it in numbers, your knowledge is of a meager and unsatisfactory kind."

The **metric** is what we are studying and **measurement** is how we study.

Measurement requires a method for how to do the measuring, a standard unit, and a calibrated gauge. Then each reading must be recorded. Each metric will dictate what level of precision is required. Too much precision can be as detrimental as too little precision.

The big distinction between error and variation is, **error** can and must be avoided. **Variation** is the natural deviation of a measurement and as our process improves the variation may be reduced.

There are four types of measurement error: operator, gauge, method, and blunder. We must guard against each error type at all times.

Chapter Three:
Charting on a Control Chart

Statistical Process Control is comprised of 1) control charts and 2) capability studies.

A control chart is the operator's tool to monitor a particular machine or system over its time of function. It includes an alarm system with a mechanism for detecting any type of shift or change. This alarm must be dynamically recalibrated to reflect the current status. The control chart supports both variable and attribute types of data. The variable control charts are $\overline{X}$ and R and Individuals.

A capability study is the management tool to assess how well the product we produce matches the product the customer desires.

To make our learning efforts easier, we will narrow our scope of interest to include only the operator and the tools that are designed to assist him. With this approach in mind, we will focus all our attention and work on covering control chart methods in detail. We will first cover the most commonly used control chart: $\overline{X}$ and R. This chapter will cover the construction details of the $\overline{X}$ and R chart.

VARIABLE DATA CONTROL CHARTS

Variable data is continuous measurement of things like temperature, length, width, diameter, and time. A set of monitors exists for monitoring this kind of situation. As we learned earlier, all three statistical components must be monitored: central tendency, variability, and distribution. Our control charts are the ongoing surveillance tools for a continuous process to alert us to any change in the statistical components. A control chart displays changes over time.

The charts then are analyzed in pairs to monitor central tendency and variability.

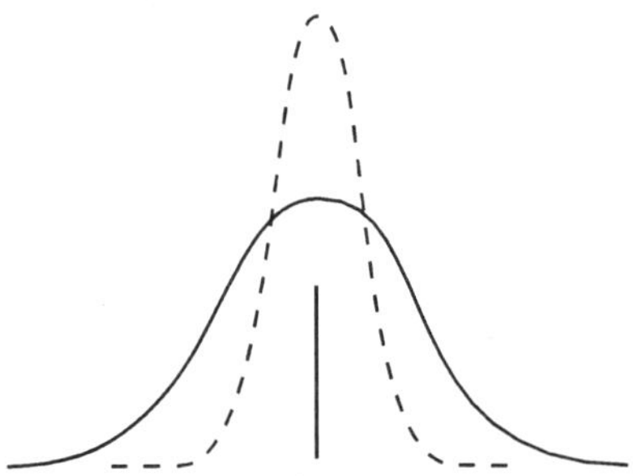

Variability Change

Variability: One chart displays changes in the variability of the measurements. The variability can either tighten or spread. The example here shows the variability collapsing. The solid line in this chart shows the original distribution. The dotted curve of our changed process is tightened because of the variability change. This picture depicts the effect of the reduced variability.

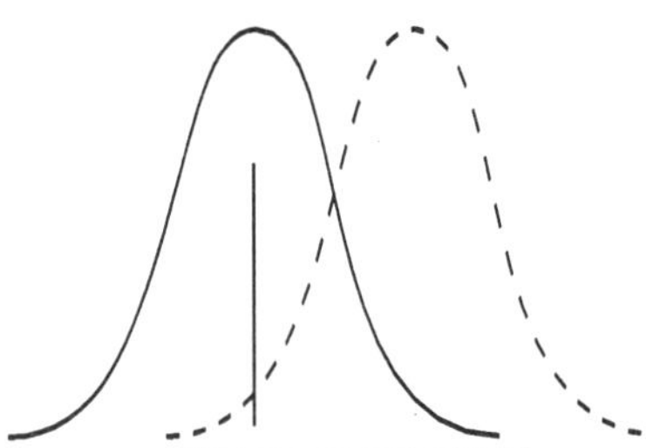

Central Tendency Change

Average: A second chart displays the average values of the samples to detect shifts in the central tendency. The solid curve represents the original distribution. The central tendency can shift left or right from its current position. In this example, the center has moved to the right from the current center position. The dotted curve shows the center point of the distribution changing to a higher level.

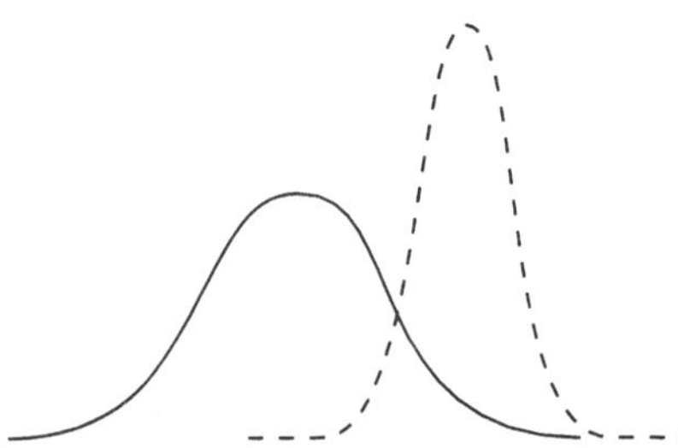

Both Central Tendency and Variability Changing

Shifts in both the central tendency and variability can occur, and each can occur by itself. Control charts must be able to detect both the central tendency shift and the variability shift.

Now, we see in the graph that shifts in both central tendency and variability can occur simultaneously as well. If only one control chart attempted to detect two simultaneous shifts, the alarm might not draw attention to both areas.

The reason for having two smoke detectors, one in the kitchen and one in the bedroom, is to help reduce the search area for the fire causing the alarm. If the smoke detector in the kitchen is buzzing, then our first investigation is going to be in the kitchen. No one would start looking for the fire at the end of the house opposite from the smoke detector that is sounding. This reduces the area that we must search and investigate. We can quickly go to the area of the house where the alarm has sounded. Now our fire search is focused in one specific area to determine the cause and what appropriate action will be needed. It is much easier to determine the proper action to take when our efforts have been focused.

In like manner, we always need two control charts, one to monitor variability and one to monitor central tendency. This, like the multiple smoke detectors, will help us reduce our search and investigation area. Now let's look at the pairs of charts that are classically used to monitor variable data.

THE TRADITIONAL VARIABLE CONTROL CHARTS

In the 1920s, Dr. W. A. Skewhart developed a series of control charts to monitor variable data. Each set was designed to monitor for a particular situation. The classic charts are $\overline{X}$ and R and Individual charts.

$\overline{X}$ and R are the prime charts for monitoring a high volume continuous manufacturing facility. The Individuals chart is for times when subgrouping is impossible or undesirable.

All of the classic control charts are actually comprised of two control charts. One chart monitors central tendency for any changes. The other control chart monitors for changes in the variability of the process. The following table shows what each chart is designed to monitor:

Chart	Central Tendency	Variability	Distribution
$\overline{X}$ and R	$\overline{X}$	R	Central Limit Theorem
Individual	X	Moving Range	Histogram

Remember, do not confuse control charts and control limits with a management chart to assess customer specifications. Control limits help us determine if a process is producing consistent output and not changing. Later, our strategic decision-makers will apply specification limits to the distribution of a controlled process to determine its capability.

A STORY FOR OUR LESSONS

We will need a setting and data to use to help explain the control chart lessons. Since I love chocolate candy, we will use candy making as our example.

As a boy growing up in Nashville, Tennessee, my greatest treat was a candy called GooGoo® Clusters. These little bits of heaven were, and still are, made in Nashville by Standard Candy Company. A GooGoo® Cluster contains the finest ingredients known to man. In the center, caramel and the best peanuts grown in southern Alabama and Georgia surround the finest marshmallow. The best chocolate produced in Tennessee, or anywhere else in the world, covers the entire confection. I am a connoisseur of these candies, and being a connoisseur, I know that a GooGoo® Cluster weighs roughly 50 grams because, of course, every time I eat a GooGoo® Cluster, I weigh and inspect the package.

My grand ambition and most exciting fantasy is to own a candy company. Then, I could make my own chocolate-peanut clusters. I could eat as many as I want, I could give them to my friends, and I could sell them. We will pretend that, because of my boyhood GooGoo® influence, I have purchased a chocolate-peanut cluster candy company. We will use this childhood fantasy to explain the operational tools required to run a facility.

In my fantasy, I purchase a candy company. I decide to build a new state of the art plant in Nashville, Tennessee. I bring in the most competent engineers in the world to design this plant. We hire the best construction company, and we use the best materials and equipment. Upon completion, the plant will make clusters that weigh about 50 grams.

First we will analyze the viscosity of the chocolate. We will start by looking at Individual charts.

POSTING TO AN INDIVIDUAL CHART

Individual charts are used when we are able to take only one sample. Individual charts are also used when multiple tests in a subgroup yield the exact same reading. Examples of this situation would include chemical processes, expensive products, low volumes of product, or homogeneous readings. The individual control charts are comprised of an Individual chart for monitoring central tendency and a Moving Range chart for monitoring variability.

Individual Control Chart, sometimes referred to as the X chart, is designed to monitor the movement or changes in the central tendency of the process. The X chart is comprised of the following plot lines:

Monitor: $\overline{X}$ is the benchmark of the central tendency of the process. The upper control limit is where no measurements are expected above. The lower control limit is where no measurements are expected below.

Data: X is the data point to be monitored.

This is an example of the Individual or X control chart.

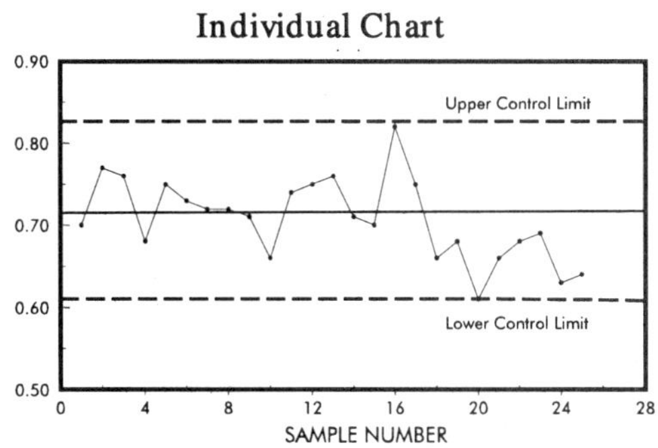

Moving Range Chart is designed to monitor the movement or changes in the variability of the process. The moving range chart is comprised of the following plot lines:

Monitor: $\overline{R}$ is the benchmark of the variability of the process. The upper control limit is where no R-values are expected above. The lower control limit is where no R-values are below.

Data: R is the moving range between the current and last reading and becomes the data point to be monitored.

This is an example of a Moving Range control chart.

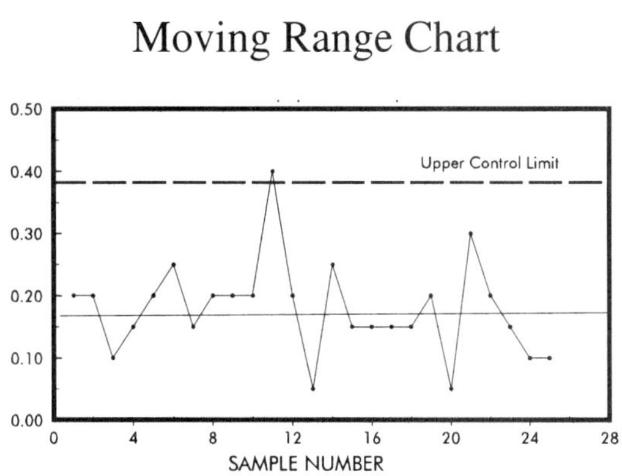

X CHART EXAMPLE

The data for the X chart of the Individual chart will be the individual values. For the variability of the data a moving range will be used which compares the current and prior value to compute the distance between the values.

The table below shows the 1:00, 2:00, 3:00, 4:00, and 5:00 readings. Our candy plant has a chocolate vat. Every hour the chocolate viscosity is measured.

Subgroup Sample	1:00	2:00	3:00	4:00	5:00	6:00
1	10	12	14	15	13	

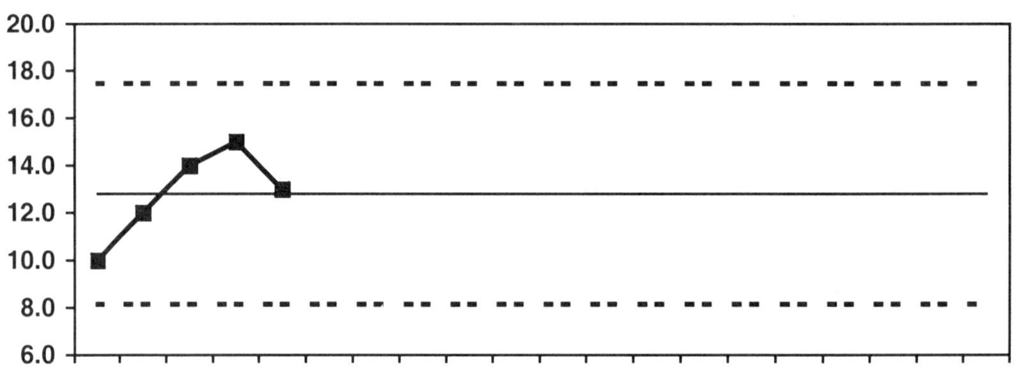

The moving range value is now computed for each time period for monitoring the variability of the values. The 1:00 reading has no prior reading so no moving range can be computed. Our first moving range is computed for the 2:00 reading, where the 2:00 data of 12 is compared to 1:00 reading of ten. The moving range at 2:00 is two; it is plotted on the Moving Range chart. The formula for the 2:00 moving range is shown below:

$$R = \left| X_{Current} - X_{Prior} \right|$$

$$= \left|12 - 10\right|$$

$$= 2$$

Notice that the absolute value or positive value of the range is always used. This keeps the distance, whether moving up or down, relative. The table below shows the individual data and the moving range for all the example data.

Subgroup Sample	**1:00**	**2:00**	**3:00**	**4:00**	**5:00**	**6:00**
1	10	12	14	15	13	
R		2	2	1	2	

These readings will be plotted on the Individual chart below as data.

The following chart shows our solved problem:

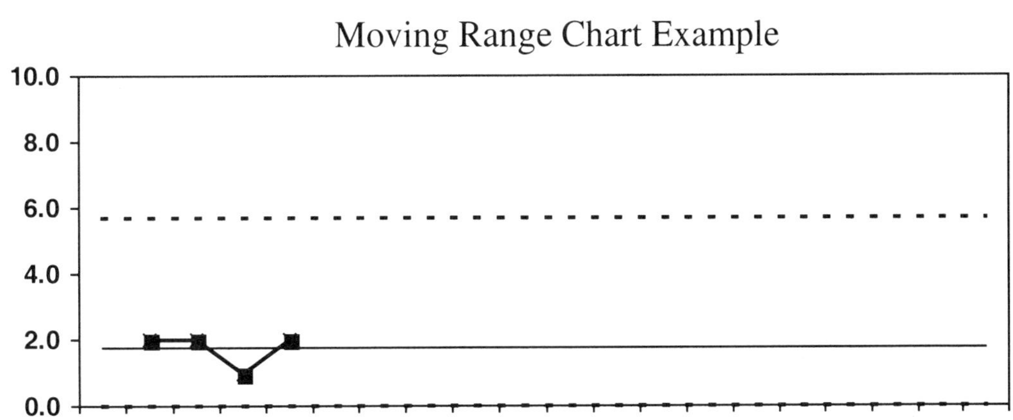

Posting to the $\overline{X}$ and R chart

This chart, the most commonly used, is geared to high volume manufacturing. Users typically monitor product metrics such as the weight of a biscuit, the diameter of a drinking glass, or the weight of a candy bar. This chart typically uses samples of five in a subgroup. Be sure that you plan for adequate resources to handle this large quantity of data. The $\overline{X}$ and R are normally used to monitor a high volume discrete product.

$\overline{X}$ chart is designed to monitor the movement or changes in the central tendency of the process. Throughout this book the term $\overline{X}$ will be used for the average of the subgroup and $\overline{\overline{X}}$ for the average of the averages. The $\overline{X}$ chart has the plot lines for data points and alarms.

This chart is an example of an $\overline{X}$ chart.

$\overline{X}$ Chart

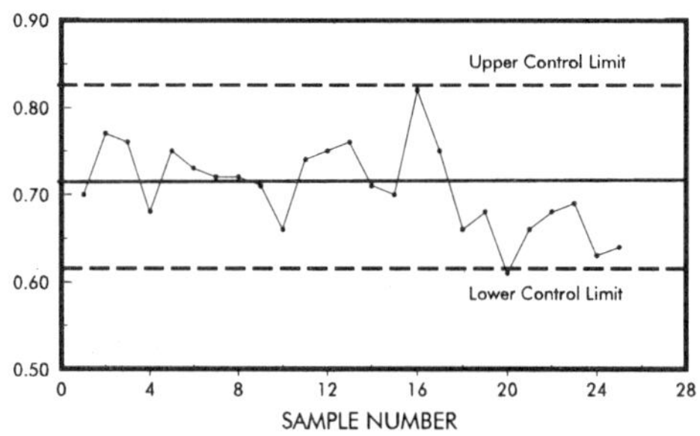

R chart is designed to monitor the movement or changes in the variability of the process. The R chart has the plot lines for data points and alarms. This chart is an example of an R chart.

R Chart

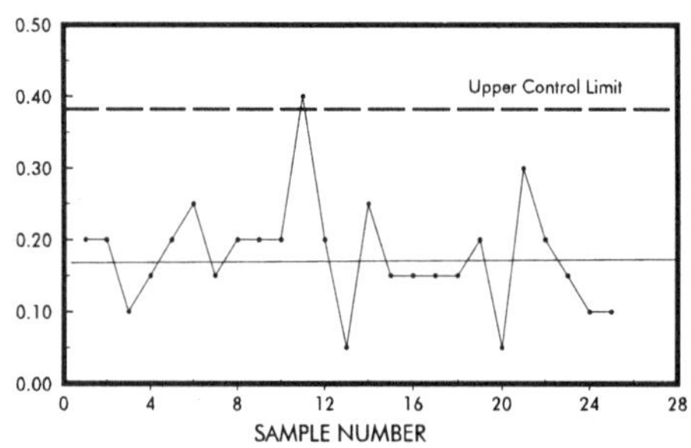

The $\overline{X}$ chart and R chart form a powerful team with the $\overline{X}$ chart monitoring central tendency and the R chart monitoring variability. The prime objective is detection of minor shifts or changes while the changes are still small and manageable. Now let's learn how to construct both charts.

For a simple example, I have prepared the following product metric measurement data to analyze. Our manufacturing facility for this exercise is a high volume candy plant. Candy is coming from one machine down a conveyor belt in the sequence that it is produced. Each set of readings is taken consecutively and the subgroups are taken every hour. We have test data for the 1:00, 2:00, 3:00, and 4:00 time periods. This data will be used to generate our initial control limits.

This volume of data is only for this example and would not be sufficient volume to build a good set of limits. In the real world and in actual practice, larger volumes are required to provide a vivid picture of the process. As a rule of thumb, thirty subgroups would be required to determine the initial control limits.

Subgroup	**1:00**	**2:00**	**3:00**	**4:00**	**5:00**	**6:00**
Sample						
1	10	14	10	10		
2	11	8	10	11		
3	13	10	9	9		
4	9	10	10	10		
5	10	9	10	10		

$\overline{X}$ AND R CONTROL CHART DATA

The following are formulas to compute the 1:00 subgroup control chart data points. They are used to derive the data to plot for the $\overline{X}$ and R charts. Average data is used to provide a means to alert operators when a shift in the process has occurred. This should eliminate the potential for over adjusting the machine because of a false alarm.

X are the data readings for the particular subgroup where N equals the size of the subgroup. $\overline{X}$ is the average of the subgroup while R is the range for the subgroup.

$$\overline{X} = \frac{X_1 + X_2 + X_3 + \ldots + X_N}{N}$$

where N equals the number of samples in the subgroup.

$$R = \left| X_{highest} - X_{lowest} \right|$$

Look in-depth at the 1:00 readings of 10, 11, 13, 9, and 10. These are taken at precisely 1:00 and each reading is from the next five consecutive candy clusters. The actual readings are fed into the control chart mechanism. The data for the control chart is the normalized values or averages. The $\overline{X}$ value is the average of the five readings. The R-value is the difference between the highest of the five readings (13) and the lowest of the five readings (9).

The 1:00 $\overline{X}$ control chart reading is computed as follows:

$$\overline{X}_{1:00} = \frac{10 + 11 + 13 + 9 + 10}{5}$$

$$= \frac{53}{5}$$

$$= 10.6$$

The 1:00 R control chart reading is completed as follows:

$$R_{1:00} = 13\text{-}9$$

$$R_{1:00} = 4$$

Now each one of the hourly subgroups must be calculated in the same manner. The results of these calculations are shown below:

Subgroup	1:00	2:00	3:00	4:00	5:00	6:00
Sample						
1	10	14	10	10		
2	11	8	10	11		
3	13	10	9	9		
4	9	10	10	10		
5	10	9	10	10		
Total	53	51	49	50		
Samples	5	5	5	5		
$\overline{X}$ = Total / Samples	10.6	10.2	9.8	10		
R	4	6	1	2		

On a control chart, the actual individual candy readings will not be used. From this point on in this chapter, no use of the individual readings will be made. Therefore in the tables that follow, all of the individual readings have been eliminated.

Subgroup	1:00	2:00	3:00	4:00	5:00	6:00
$\overline{X}$	10.6	10.2	9.8	10.0		
R	4	6	1	2		

Now we can take these values and use them in the calculation of our control limits.

$\overline{X}$ AND R CHART EXAMPLES PLOTTED

The $\overline{X}$ chart is designed to monitor the movement or changes in the central tendency of the process. The $\overline{X}$ chart from our simple example is shown below:

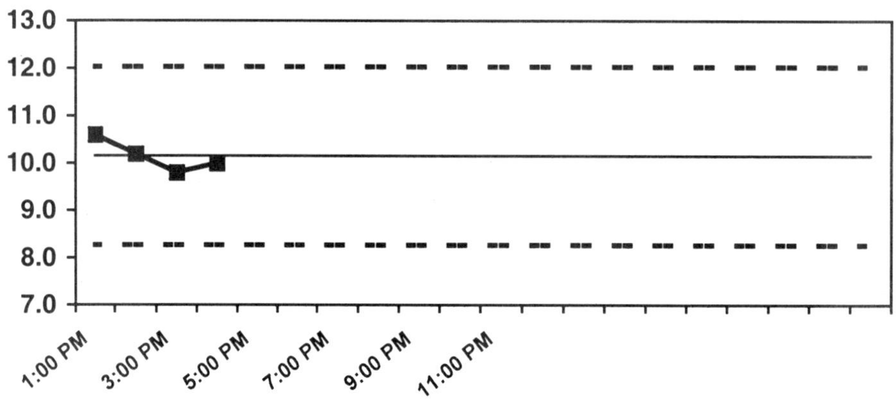

The R chart is designed to monitor the movement or changes in the variability of the process. The R chart from our simple example is shown below:

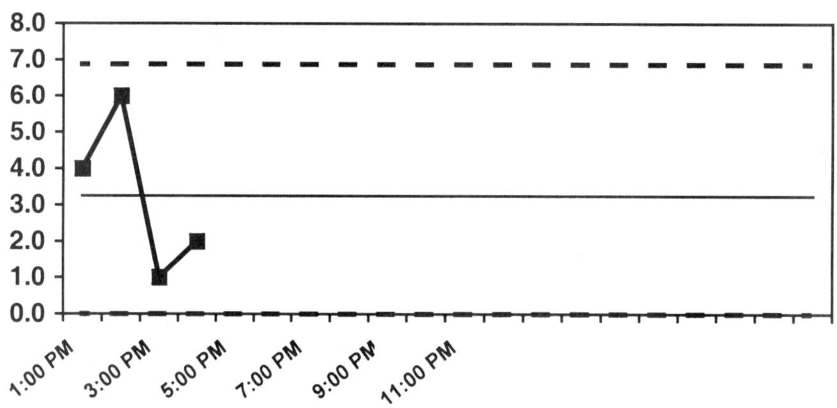

CAUTION! Once the process is in control, stop recalculating the control limits. The recalculation of control limits is a joint decision of both strategic and tactical decision-makers. Never let computers automatically recompute your control limits without direct human decisions.

$\overline{X}$ AND R CHART SAMPLE PROBLEM

The Candy Plant Revisited

For our exercise, the candy plant is revisited and this time the metric that we are monitoring is the weight of our candy. We want this target weight to be 50 grams. In the table below, readings are taken every hour. The hourly readings are pulled from the conveyor belt in a subgroup of five chocolate candy clusters. These individual chocolate clusters are pulled from the conveyor in consecutive order with the objective of better understanding the weight of the candy as it is produced.

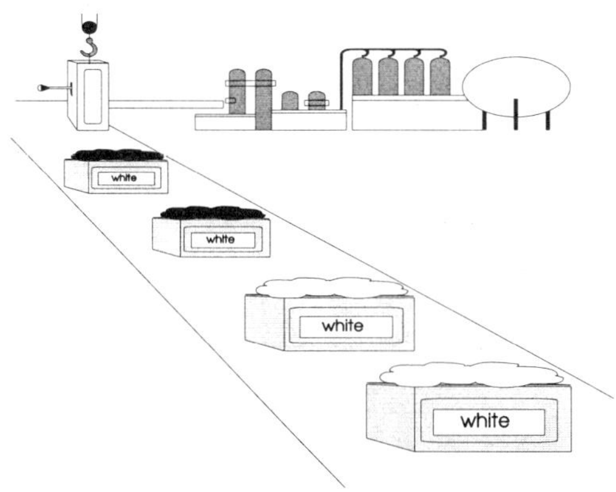

EXERCISES FOR $\overline{X}$ AND R CONTROL CHARTS

The $\overline{X}$ control chart detects shifts in central tendency. Does our data have any shifts? To answer this question, calculate the $\overline{X}$ and R-values for our candy machine and write the answer in the table below.

	1:00	2:00	3:00	4:00	5:00	6:00	7:00	8:00	9:00	10:00
	50	55	60	45	40	50	55	60	45	40
	51	56	61	46	41	51	56	61	46	41
	52	57	62	47	42	52	57	62	47	42
	49	54	59	44	39	49	54	59	44	39
	48	53	58	43	38	48	53	58	43	38
$\overline{X}$										
R										

Use the chart below to plot the $\overline{X}$ data points. The control limits have been calculated and posted on the chart. Post your values and take note of how the chart looks.

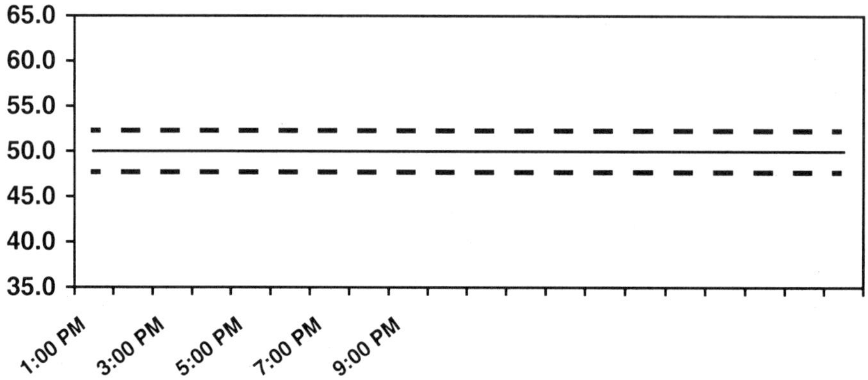

In the next chapter we will explain the control chart alarms.

Now plot the R-values.

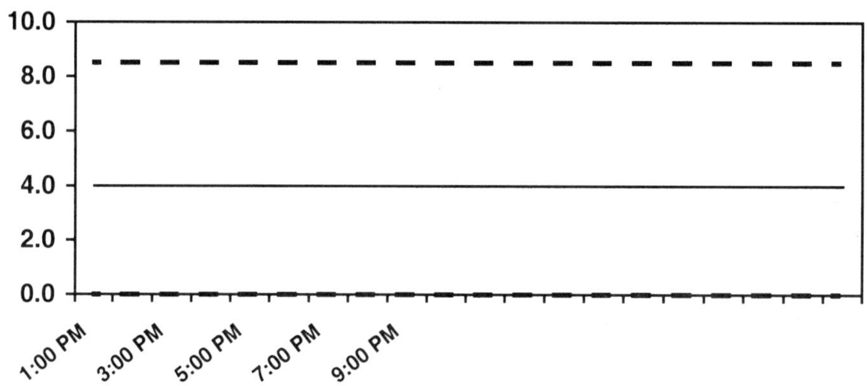

Does our computed data show any shifts? These completed control charts show us the answer to that question. Check Appendix B if you are unsure of your answers. We will see what the alarms are and how to detect a change in the next chapter.

The formula and constant tables for computing the control limits can be found in the Appendix of this book. For a more rigorous explanation of these charts, their calculations, and the theory and concepts behind these charts, please refer to ***SPC**, third edition.*

Chapter Four:

Reading a Control Chart

CHANGE DETECTION

The overall objective of control chart methods is to detect product or process changes. All control charts use the same set of alarms to detect change. The term "out of control" simply means a change has been detected. The demonstrated effectiveness of a control chart is attributable to its objective and systematic analytical methods. Control charts are tactical tools to detect change in a continuous operation. A series of easy to use and read alarms are provided for the tactician (operator). These alarms are simple but effective.

WHEN A CHANGE IS REAL

Often the natural ebb and flow of a process is mistaken for real change. With these false changes, operators begin to adjust the machine incorrectly. These incorrect adjustments lead to more false change alarms and, thus, more incorrect adjustments.

To prevent that misinterpretation, a set of benchmarks is required to establish what is typical for a particular machine. The control chart's limits provide the means to differentiate between what is typical and what is unusual behavior for a process and product. Remember, what is usual for one machine may be bizarre for another. SPC provides the tools to separate the typical from the unusual.

The SPC tool of control charts is comprised of a few simple alarms that note a change is taking place and that an immediate initiation of the causal search should be advanced at full speed.

We are narrowing our scope to focus our attention on the tactical issue of consistency. Our data is gathered and each set is then immediately tested for a shift. If no shift is detected, no action or adjustments are necessary. Only when a real change is detected is research required. Since our tactician's (operator's) time is so precious, he must only research a real change. He must not waste his time researching what is the natural ebb and flow of a process or product.

Remember; never take action just because an alarm is noted. The cause must always be found before taking action. Earlier we discussed the smoke detector. We would never bring a garden hose into the house to put out a fire without an investigation. If the fire turns out to be a grease fire, the water will simply make the situation worse by spreading the fire. The cause must be identified before any action is taken.

STEPS FOR INTERPRETING AN ALARM

There is a control chart for variability and another for central tendency. Variability is often skipped, misunderstood, and ignored. Thus, I always monitor the variability first, and check central tendency second. Once each component has been monitored for change, I look at both charts together.

To summarize, these three steps are as follows:

1. Check Variability
2. Check Central Tendency
3. Check Both – variability and central tendency together

We must assure that we always check both variability and central tendency for change. The most common mistake in controlling a process is to test only for shifts in the central tendency, or center; variability must be monitored as well.

Understanding exactly what the alarms are is crucial. The next few pages show the alarms used on a control chart. Remember that these alarms are simple to detect change.

SINGLE POINT OUTSIDE CONTROL LIMITS

Any data point outside control limits requires investigation because the odds are very remote that a controlled process could produce such a measurement by accident. The chart should trigger an alert to investigate and determine the cause and the impact of the change.

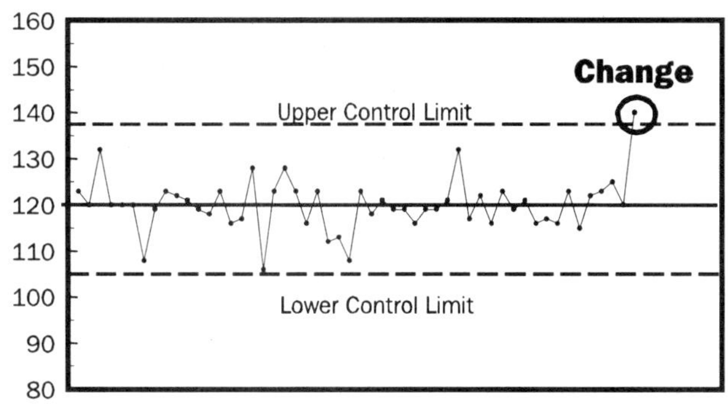

Any point falling outside the chart control limits is an indication that a change is occurring. Any range value falling outside the chart control limits is an indication that variability is changing. An immediate investigation of either situation is required.

Remember, as we stated before, never act on the alarm. Before any action can be taken, the cause of the alarm must be identified. The control chart does not tell us the cause but only alerts us to start the investigation. The quicker our investigation starts the better our chances that we will determine the cause.

A SERIES OF POINTS OUTSIDE OF THE CONTROL LIMITS

Since the odds of a single data point outside the control limits is rare, then a **series** of data points outside the limits must be extremely rare.

The probability of having multiple consecutive data points outside the control limits is low. It would be almost impossible to have four consecutive data points outside the control limits that did not show change. This leads us to the conclusion that this must truly be a shift in the metric that is being monitored. The control chart below shows this particular situation.

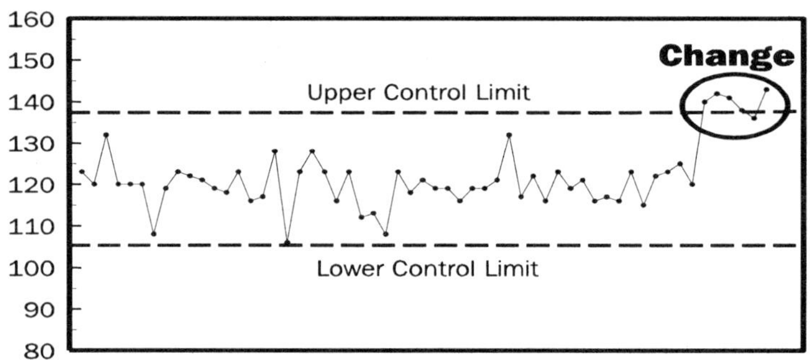

THE "JUMP IN LEVEL" ALARM

Many people mistakenly look only for the alarms that we just covered. Control charts were designed to detect much more subtle changes than those. Because control chart methods are designed to detect minor changes, many of our alarms are based around values that are inside the control limits. These alarms are geared to detect the most minor of shifts for any one particular machine. The first change that is detected with data inside the control limits is a **jump in level**. This particular alarm was geared for minor shifts above the mean of the central tendency.

The alarm is seven consecutive data points above the center point of the control chart, with none of the data points falling outside the control limits. The chart below reflects this particular alarm.

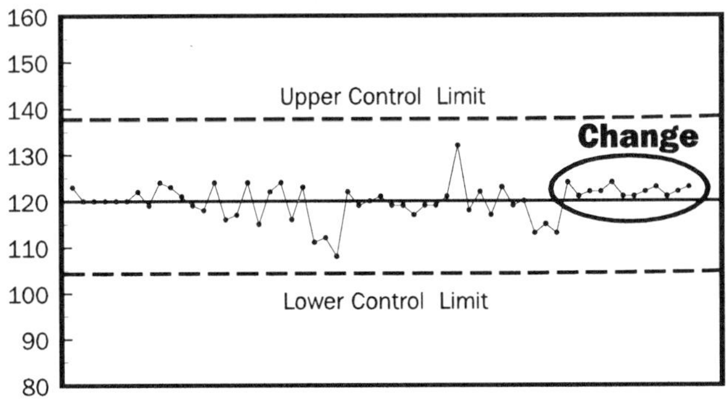

An alarm similar to the above is seven consecutive data points all below the center point of the control chart, but none of the data points are outside the control limits.

The first indication of an alarm is when seven consecutive points are on the same side of the centerline but still within the control limits. The more consecutive values on one side of the center, the bigger the alarm.

All the alarms are alerts that a change has occurred in the process or product. Never stop the machine because of an alarm on the control chart. This alarm may be a signal of an improvement!

THE "TREND" ALARM

Another minor shift that is important is a trend. Trends can take the form of going up or coming down. Once again, minor shifts are what the control chart's detection mechanism looks for. A trend can also be found where the data is inside the control limits.

This particular alarm, illustrated below, is geared to detect a slow shift in a downward direction ("trending down"). It is described as seven consecutive points with each point lower than the last point and no points outside the control limits.

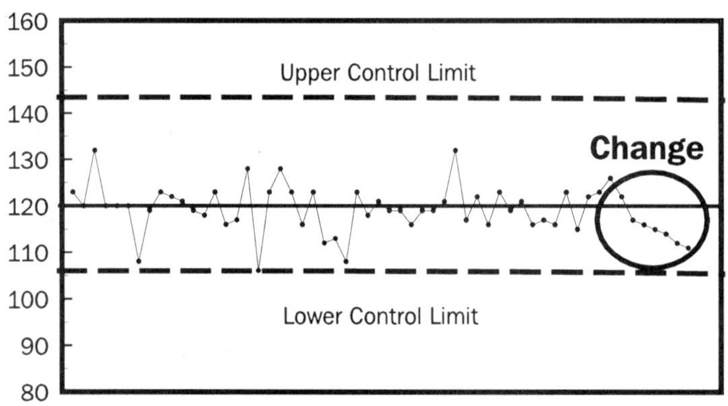

The same idea applies to the reverse, where the trend is increasing. This particular alarm is geared to detect a slow shift in an upward direction ("trending up"). This alarm is seven consecutive points with each point higher than the last point and none of the points outside the control limits.

Keep in mind, control charts must be viewed in a total context of all the information that is available. Just because no control chart alarms have gone off does not mean that other information might not cause an alert. As an example, if a drill bit is dull and needs replacing the information about the bit should tell us to change the bit even though no control chart alarms have sounded.

THE NORMAL DISTRIBUTION

Everyone has been graded on a bell shaped curve at some point in school. The name for this bell curve is the normal distribution. The normal distribution is used to convert the actual grades or test scores to a letter grade.

The normal distribution is symmetric around its average. Fifty percent of the data points are above the average and fifty percent of the measurements are below the average. The frequency of occurrence (or the number of occurrences) drops rapidly as we move further and further away from the center. This characteristic is true both above and below the average. The frequencies of occurrences above and below the average are symmetric.

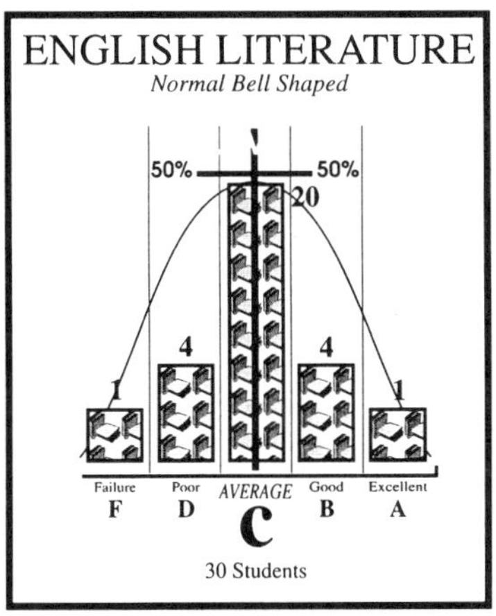

The chart shows the spread of grades from an English literature class of thirty students. This chart shows how a typical class in English literature could have been graded using a bell shaped curve. Out of a class of thirty students twenty would receive a "C", four would receive a "B", another four would receive a "D", one would receive an "A", and one would receive an "F". This would represent a typical class.

THE "TOO CLOSE" ALARM

If the class grades were not spread like the above but every student received a "C", we would immediately conclude that something was wrong in this class. If all the past classes had shown the normal bell-shaped spread and this new class had all "Cs," we could now conclude that something about this current class had changed. The too close alarm is analogous to all the students receiving a "C" grade.

The zone one area on a control chart includes 68% of the values and is shown below. If all of our measurement chart postings fall within this zone one (all "Cs"), we can conclude that something about the process has changed. If too many of the values are clustered around the mean, an alarm is indicated. Remember, this is not an indication of bad product or good product, but an indication of change that needs to be investigated.

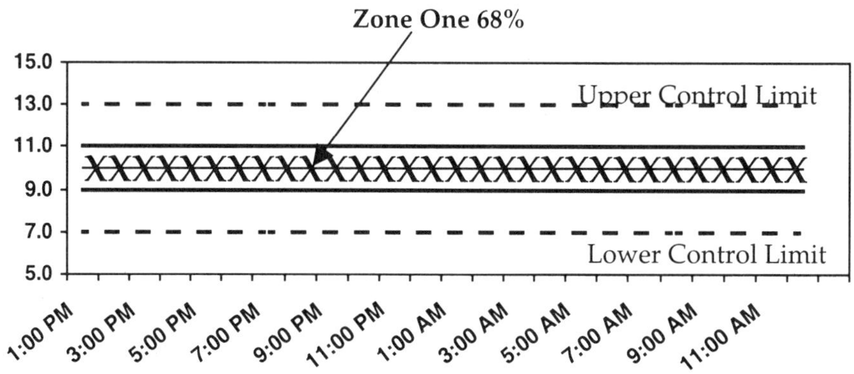

On the control chart below, all the data points are inside the zone one area. This chart is an example of the "too close" alarm going off. This alarm like all the other alarms must be investigated immediately.

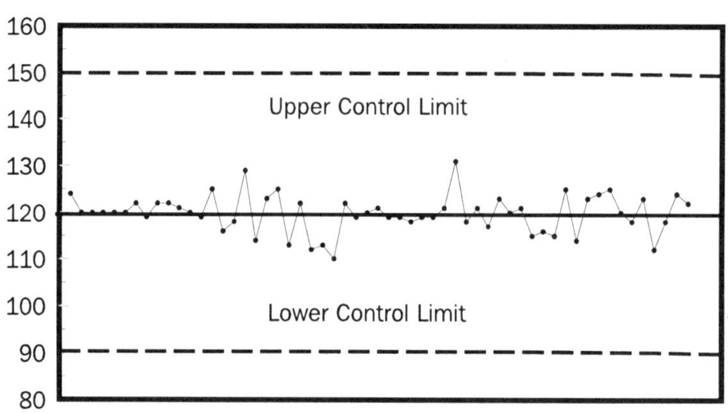

Minor change going unnoticed over time may seem trivial, however these minor changes have a compounding effect. Over time, they allow us to miss major improvement opportunities or slide into major defect problems.

THE "TOO FAR APART" ALARM

In the bell-shaped English literature class, four students would be expected to make a "B" and another four students would be expected to make a "D." Zone two shown below on the control chart is like the "B" and "D" of the literature class and includes about 27% of the values.

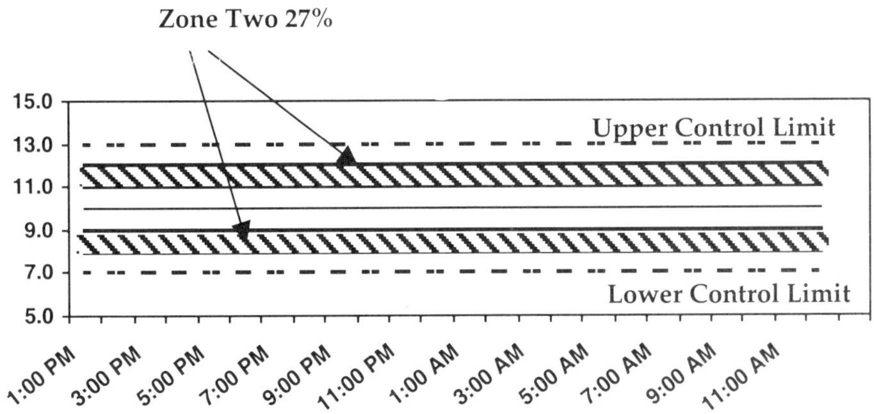

One "A" and one "F" would be expected. Zone three, shown below on the control chart, is like the "A" and "F" of the literature class and includes about 5% of the values.

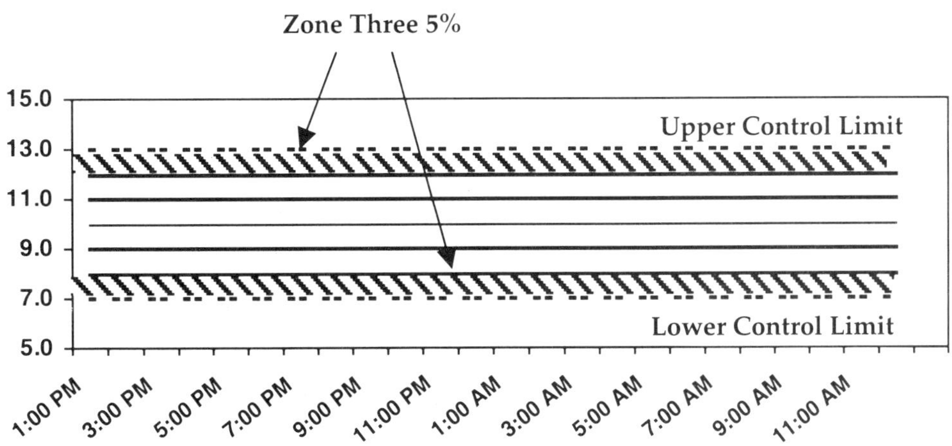

THE SPREAD ALARM

The total number of data points in zones two and three should be 32%. The alarm here is if more than 32% of the data points are in zones two and three. This becomes just as big an alarm as all the prior alarms. If everyone in the literature class got As, Bs, Ds, and Fs, but no Cs, then we would surmise that a change had occurred.

On the control chart below, none of the values fall inside zone one (no "Cs"), but no values are outside the control limits. This is an example of the spread alarm.

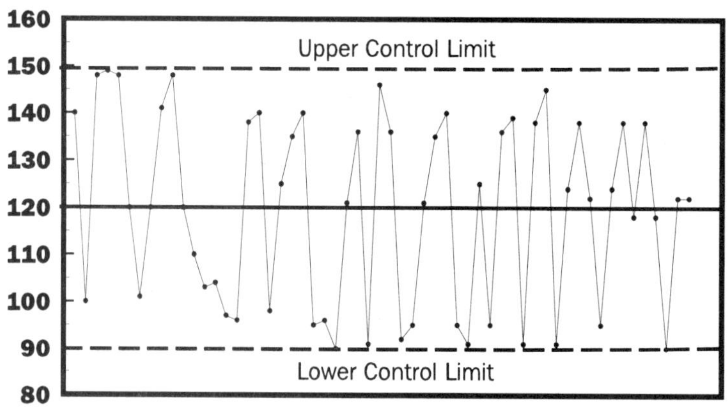

SUMMARY OF INTERPRETATION GUIDELINES

The table below summarizes the alarms that we use to monitor or detect any shifts, patterns, or trends.

- Single point outside the control limits
- A series of points outside the control limits
- Seven consecutive points increasing (or decreasing) from the last point
- Seven consecutive points above (or below) the average
- More than 68% of the points in the first zone
- More than 32% of the points outside the first zone but still within the control limits

Chapter Five:
Using a Control Chart

THE USE OF CONTROL CHARTS

The control chart component is one player on the SPC team. A control chart helps the operator understand his machine and establish how to correctly and consistently run it. The chart also establishes the typical patterns for the machine and alerts the operator to any deviations from the norm. These deviations are "out of control." Many people are confused by the term **out of control** but it simply is another word for change.

If no change is detected, we must wait for the next readings. If a change is detected, our investigation begins to determine the cause and assess the impact. We decide what action to take once all the prior steps are complete. Once a change is detected, we are just at the start of our journey; we must move on to our investigation efforts.

INITIATE INVESTIGATION

Our investigation is comprised of two parts: one tactical and one strategic. A clear understanding of the roles of both the strategic and tactical players is key. Once the individual roles are clear, the players must function together as a team for effective action.

The tactical decision-makers (operators) begin investigating to determine the cause of the alarm. These causes will increase our understanding of the process and product. This knowledge is crucial for improvement. Once the knowledge has been learned, it must be retained and shared with all associates. The tacticians must also alert the strategic decision-makers that a change has occurred so that an assessment of the impact of the change can be made.

The strategic decision-makers must simultaneously (while the cause is being investigated) assess the impact of product or process change on the customer.

Once the cause has been identified and the capability determined, a course of action must be chosen. This is a team decision by both the strategic and tactical decision-makers.

We are narrowing our scope to focus our attention on the tactical issue of consistency. The following flow is how we should use our control charts. Our data is gathered and each set is then immediately tested for a shift. If no shift is detected, no action or adjustments are necessary. Only when a real change is detected is research required. Since our tactician's (operator's) time is so precious, the tactician must only research a real change. He must not waste his time researching what is the natural ebb and flow of a process or product. We are following a method of "management by change."

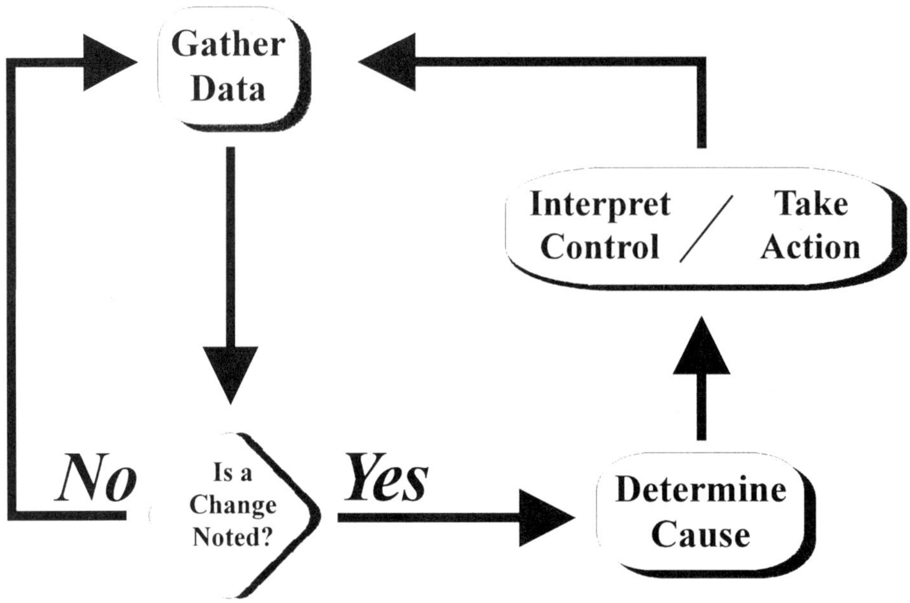

The above chart depicts the steps to action based on a control chart alarm. We regularly gather our data and test for change. If no change is detected, then the tactician continues his work until the next reading is taken. If a change is detected, the tactician begins the research for determining the cause of the change. After the change is determined, we are ready to begin the decision of how to act.

Remember, never take action just because an alarm is noted. The cause must be identified before any action is taken.

WHAT IS A CONTROL CHART ALARM

The control chart alarm is a change detection mechanism. These process or product changes are outside the typical behavior of the machine. Remember that change is not always bad; it can be either improvement or deterioration. For progress (improvement) to be made, change has to occur.

The control chart is the tactical tool to identify when the change warrants our investigation. Our investigation is an attempt to determine the reason the change took place. This is referred to as finding the cause. As we find the cause or causes our knowledge will grow, and with knowledge comes improvement.

In many ways the control chart (when used in conjunction with strategic tools) is our improvement detector. Actually the control chart can only find change, but without the control chart, change will go unnoticed. This is particularly true of minor change. Over time, many small, minor changes will accumulate to major change.

Once a change has been documented as more than just a fluke, the tactical role becomes even more focused on the correct and consistent running of the machine. The tactical operator must now focus on what caused the change while the strategic focus is to assess the impact. What is a cause and how do they work?

WHAT ARE CAUSES

Knowledge base linking is the attempt to determine what caused something to happen. Effects are the things that happened. Deducing what caused the effect is the most difficult task we will encounter. Since all of the facts are never known, determining the cause of an effect is often difficult. True cause identification is an ongoing effort that requires a vast amount of analytical experience. It will require much time, analysis, thought, and effort. We will continually add to our true causes, and in doing so, we will increase our knowledge base.

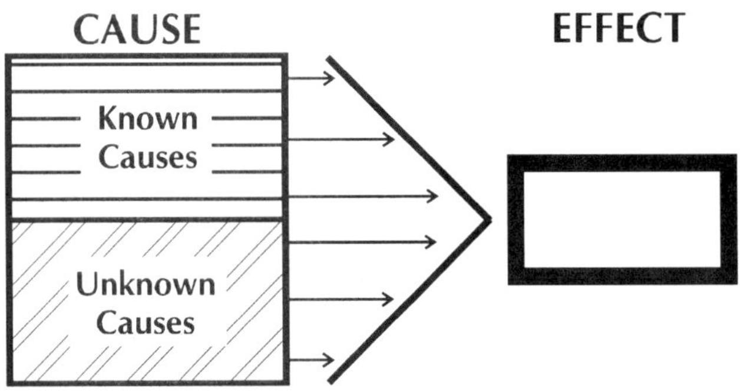

Since we never know all the facts or have a complete picture, we must continually review our determinations and findings. We must never view our determinations as laws or standards that cannot be improved. The disparity of known causes directly impacts our determinations and is due to a lack of all the facts or a failure to understand the meaning of the facts we have. The magnitude of this disparity must be continually investigated.

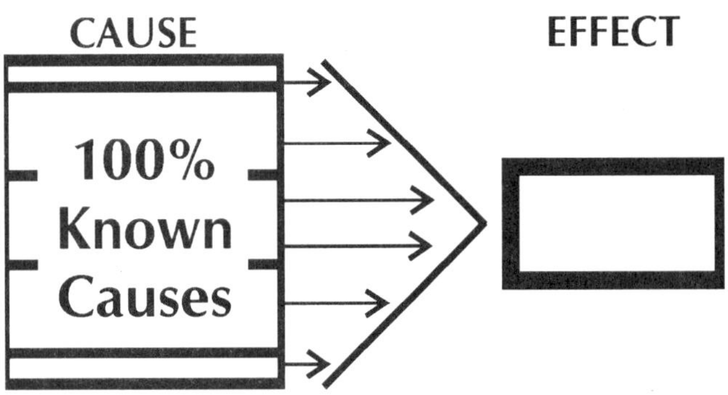

The Ultimate Goal

WHO IS BEST EQUIPPED TO FIND THE CAUSE

We first think of the machine designer, product designer, or process engineer as the most likely candidate to determine the cause of a change. These individuals are well equipped with technical skills. However, the investigation process must begin quickly so that the causal identification trail does not get cold. The problem with all the people listed is that their timely availability is questionable. Engineers, management, and designers are strategic and must be involved in the final action plan.

Change happens at all times of the day, weekends, holidays, and even at 3:00 AM on Saturday. Since our efforts must follow the *Church of What's Happening Now* philosophy, the person closest to the machine must be the causal searcher. This narrows our scope to the prime person most familiar with the machine - the operator. The operator, our tactician, will lead the efforts to finding the cause of the change.

All the technical resources that can support the operator must be brought into play to find the cause. As the operator finds these causes, the knowledge base about the process and product will be increased.

Our objective is to continually reduce this disparity and achieve total knowledge of the true causes. Over time we must strive to increase our knowledge base.

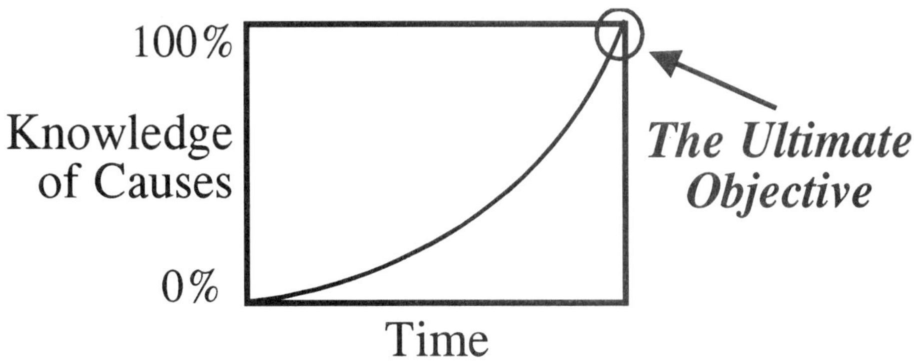

Effects are rarely the result of a single simple cause. Since most causes are seldom known, we should expect gaps in our knowledge base. These gaps of knowledge should keep us alert to the fact that effects are not always the result of known causes. Serious trouble will be encountered if the facts are forcibly stretched to explain the effect. A wise person is one who appreciates his lack of total knowledge.

To build our total knowledge base, our analysis must be based only on a historical record. The historical records allow us to begin to build a causal linking. This will not preclude our adding knowledge of changes in the future to better understand the cause. This is like driving a car. We look through the rear view mirror to understand where we have been. This view also allows us to see cars coming from the rear. We look through the front windshield to understand things coming up such as a curve in the road or a car on our side of the street.

The data for our analysis must begin at the basic element. Our analysis must begin both from the bottom and from the top, allowing the art of causal analysis to take place.

Our causal linking is easy when we view the most immediate occurrence that is known. The further removed the causes are from the effect, the more difficult the linking task. This linking must track the links of the chain from process to product. All products and processes are interrelated. This interrelationship impacts the outcome (effect).

To better understand the investigation and analysis that will be required, let's define the type of causes that can come into play.

ASSIGNABLE CAUSES

As we begin to study our information, a clear understanding of what a cause means and the types of causes which are possible is crucial to understanding what actions are required. We must first understand that we are looking for change. Change includes both improvements and deterioration. We must understand which type of change we are dealing with in order to understand our courses of action.

Assign: 1. To set aside or give out in portions or shares. 2. Attribute; To consider as resulting from, proper to, or belonging to a person or thing.

Cause: 1. The producer of an effect, result, or consequence. 2. The one, such as a person, an event, or a condition, that is responsible for an action or a result. 3. What brings about or is associated with an effect or result. A cause is an agent or condition that permits the occurrence of an effect or necessarily or invariably leads to a result.

When we put the dictionary's definitions into action, we conclude that assignable causes are those that definitely result in a particular effect. As we read the definitions, we see that all causes do not have equal impact.

As we increase our knowledge base, we must strive to include all causes no matter how small their impact. A series of minor causes can lead to a cumulative impact that is very large. Too often we are only looking for the Big Bang. We are searching for both major and minor causes, so that we can build our knowledge base to be all-inclusive.

Causes are neither good nor bad but simply a statement of fact. A cause is why something happened or changed. For us to make improvements, we must first detect change and then determine its cause. We will diligently search for all causes, both those bringing about improvement and those resulting in deterioration. This information will allow us to increase our knowledge base of the process.

The reason causes are so important is that they give us the information to know what will happen in a particular situation. Many times an associate who has been on the job for twenty or thirty years will still run to the back end of the machine to find out what has happened every time that he changes a knob. There is major difference between thirty years on the job and thirty years of experience. With thirty years of experience, we have built a huge knowledge base that will tell us what is going to happen to the process. Our experienced operator's knowledge base allows him to know what will happen to the product when he turns that knob.

TYPES OF ASSIGNABLE CAUSES

For us to effectively make a decision, to understand its impact, and to know the correct type of decision, we must know what type of cause is involved. There are three kinds of causes: 1) Special, 2) Common, and 3) Tampering.

Special Causes are causes within one's control. The person running the machine is directly involved in the creation of a special cause. Special causes are isolated to one particular machine and impact only that one machine. As I drive down the street, the car's course adjustments are caused by my movement of the steering wheel. My pressure on the accelerator determines the car's speed.

In the same way, only an operator on a shop floor can make adjustments to his machine. These could include the speed the machine runs, the temperature of the machine, and the feed of the machine. No matter how the operator might desire these settings to impact other machines, it's pretty obvious that these changes, like the car's course adjustments, will have no effect on other machines.

Since special causes are local in nature (assigned to the driver or the operator), local action is required to resolve them. The operator must resolve the effects by dealing with the cause locally.

Special causes are the operator's responsibility for resolution. Local actions are usually required to eliminate these special causes and can usually be taken by people close to the process. Tactical decisions and tactical actions are generally used to deal with special causes. The driver of the car above must make course corrections to assure the safe driving of the vehicle. Such corrections are tactical decisions.

Common Causes are causes outside one's control. These influences generally occur across the board for all processes, thus the name common. The operator is usually not the generator of common causes. If the purchasing department buys bad peanuts for our chocolate-peanut clusters, the operator is not involved in this cause of change. Common causes are things that cross the boundaries of one machine into a group. Common causes impact multiple machines, and the impact is similar on all involved. Using the driving example, I might choose to slow the speed of my car on a rainy day. This adjustment to my speed is caused by the rain, which is outside my control. Most other drivers would also slow down due to the bad weather conditions. All the cars are being impacted by the weather, and weather is common to all. No action by the drivers, either singularly or together, could change the weather.

All operators and their machines might be impacted by a shift to a new raw material such as new peanuts for our chocolate-peanut clusters. The raw material from the new source may be inferior and cause a rash of defective chocolate-peanut candy. No matter how sincere, conscientious, and diligent the operator, he cannot directly resolve the defects or change the raw material.

Since common causes are outside local control, sources outside the local area must resolve these causes. Generally this is thought of as a managerial issue because management has overall control of the facility and the actions on the system almost always require management action for correction. Strategic decisions and strategic actions are generally used to resolve common causes. Purchasing must resolve our bad peanut problem in order for good tasting candy to be manufactured. Purchasing must make strategic decisions to resolve the problem or strategically decide to accept the defective candy clusters. No amount of local action can resolve this problem.

Tampering Causes are changes that are caused by too much control. Tampering occurs when we are trying to take action on something that should not be changed. This is typically caused by our failure to understand and comprehend the Second Principle of Process Management.

SECOND PRINCIPLE OF PROCESS MANAGEMENT

Division of Labor is the framework for all aspects of decision-making. It must be clearly understood to separate the strategic and tactical decisions. Operations makes the tactical decisions of running the facility. Management makes the strategic decisions of assessing the facility's suitability for the job.

Tampering occurs when a tactician tries to make strategic decisions like trying to force his machine to make product that the machine was not designed to do. In a similar way, a strategist doing the tactician's job violates the Division of Labor. These violations are the reason for most tampering causes.

Please refer to my book, *Optimize Your Operation,* for a complete discussion of Division of Labor. There we rigorously discuss the potential violations to the Second Principle. Since tampering is generally caused by a lack of understanding of the Second Principle of Process Management (Division of Labor), we must assure that all appreciate this potential cause. Management must assure and take the responsibility that all associates understand Division of Labor.

We must research the many questions surrounding the chain of events that caused an effect. The real art of causal analysis and linking is determining which of these questions warrant an investment of the time and money required for further research. In many ways the research is like wildcatting for oil. Many causes will be confirmed. Many causes will be rejected. Often the research leads to more questions. The objective is to continue to build our knowledge base. This information can actually become a bigger asset than the manufacturing facility!

What is the cause of an event? Some people rush to explain the cause of an event as one single thing. These people talk as if the explanation for the event can be determined by simply tracing a single line from an individual action.

However, we are flirting with disaster by using overly simple reasoning. Instead, we must vigilantly watch for multiple causes and interactions in order to improve our knowledge base.

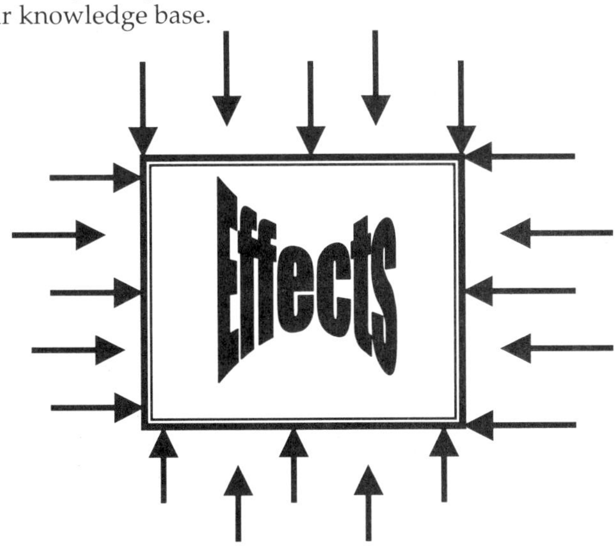

ASSESSING THE IMPACT OF THE CHANGE

Management (the strategic decision-makers) must assess the impact of the change. While the tactical work force is working to determine the cause of the change, the strategist must assess the change through a capability study.

Decisions for Action

The control chart is not a tool for action. The action must be driven from the determination of the cause and the results of the capability study.

Improvement

If the capability has improved and the cause is found, then the cause is made a standard practice and the control chart limits are recalibrated. If the capability has improved and the cause cannot be found, then the best course of action is to bring more resources into play to continue and increase our search for the cause. Otherwise, we have missed an opportunity for improvement.

Deterioration

If the capability has deteriorated and the cause is found, then the cause must be fixed and the process brought back to its original method. If the cause cannot be fixed, then the control chart limits must be recalibrated because this level of performance is now the norm for this machine.

NO CAUSE FOUND

If no cause is found, our challenge is to continually search for the cause. This situation will occur many times, but perseverance is essential to this effort. In my boyhood television days, the "Superman" episodes always talked about a never-ending search for "truth, justice, and the American way." Our search for causes must be like Superman searching for the criminal: a never-ending search for the cause of our change.

Remember, never take action just because an alarm is noted. The cause must always be found and the capability assessed before taking action.

CHAPTER SUMMARY

Special causes are within tactical control

- Tactical action is required to deal with special causes
- Should be resolved by people (tactician) closest to the process

Common causes are within strategic control

- Affects all operators and machines
- Require strategic action for resolution

Tampering causes are caused by violations of Division of Labor

- All associates must appreciate and understand their role and then execute the role

Remember, never take action just because an alarm is noted. The cause must always be found and the capability assessed before taking action.

Chapter Six:
Good Tools Misused

Many companies are frustrated because they never see results or improvement. They wonder why SPC has failed. Actually, the answer lies in they way they have used SPC.

THE REACTIONARY APPROACH TO TACTICAL DECISION-MAKING

A "reactionary" approach means that we begin our causal research only after a change in the product has been noted. Said another way, we are reacting to the problems after they occur. The problem with this approach is the added cost, the delay time, and the possibility of some bad product getting past our test method. Focusing on the product allows defects to be made, thus driving cost and delays up. This misguided approach I refer to as STATISTICAL BISCUIT CONTROL OR **SBC.**

If my grandmother had used the SBC approach, she would have waited for the biscuit control chart to alert her that the biscuits were burnt. There would be many burnt biscuits and unhappy family members as my grandmother reacted to the control chart for each batch.

In the reactionary approach to machine operation, the most important metric is the outgoing product. That means that, in our story, chocolate cluster weight determines all decisions.

The reactionary approach waits until product has shifted before any action is taken. This can lead to excessive scrap and poor decisions.

Let's use the chocolate candy factory to explain how good tools can be misused.

We set up our machine to make chocolate clusters with a target weight of 50 grams. Next, the incoming peanuts are checked and set at a density of 2.7. The conveyor speed is set to 3.1. The nut fall rate is set to 1.2 and the chocolate spray rate is set to 1.2. With each process metric set, we check and begin to monitor the chocolate cluster *weight*.

The chocolate cluster weight will drive all our tactical decisions and adjustments. Note that we will have a product shift before any action or investigation is made.

Our first day's results yield very stable results as demonstrated by our chocolate cluster weight control chart below.

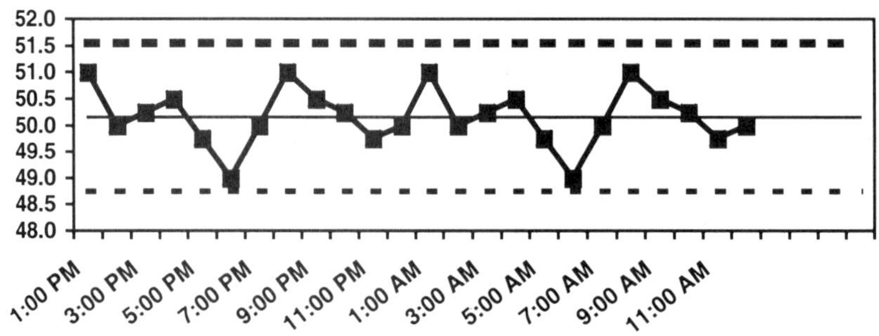

Since the machine is set to the correct target, and is running consistently, the strategist completes a capability study. The capability study shows a zero percent defect rate.

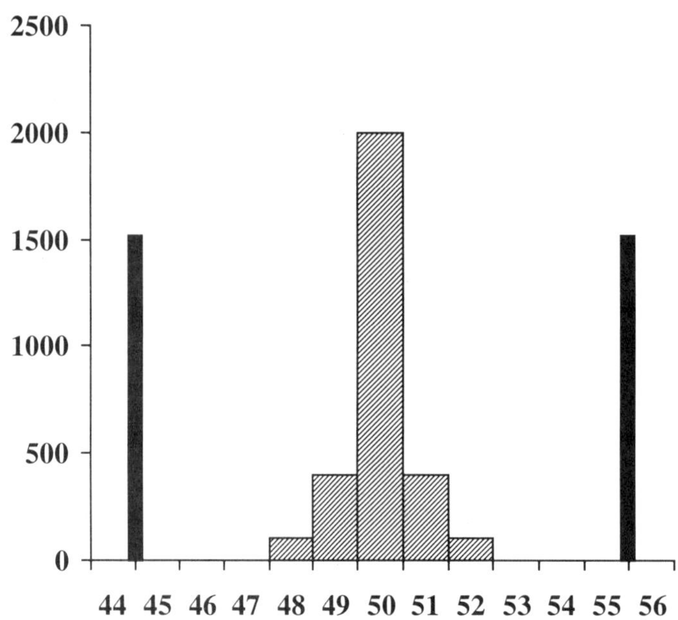

As we move into our second day, the results continue to remain consistent telling us everything is okay.

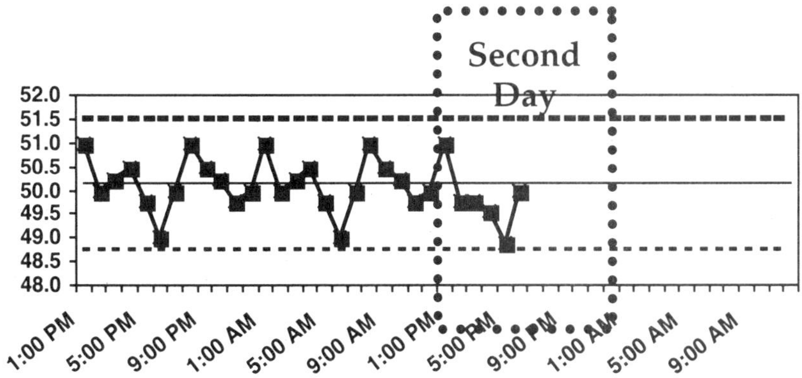

We continue to religiously measure and monitor our chocolate cluster weight readings every hour.

On the third day at 4:00 pm a chocolate weight reading of 48 grams is encountered. As we post the reading to our chart below, an outside the control limit occurrence is observed.

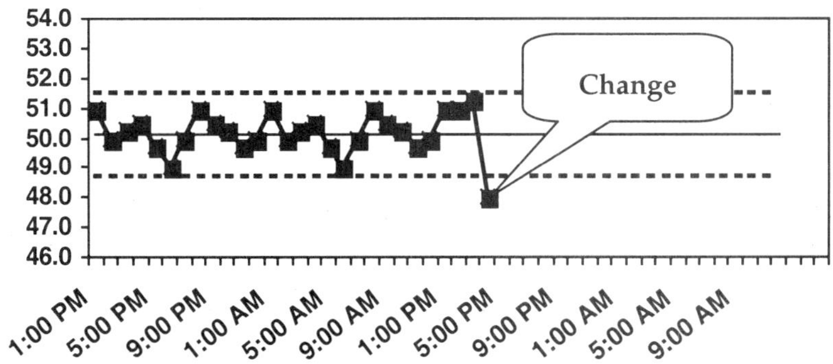

Now that a change has been identified, the operator must do two things. The operator begins to research and identify the cause of the change. He must also notify the strategic decision-makers of the change so that they can assess the impact on the quality of the product. Once the strategic process engineer has been notified, we are free to begin our research to find the cause. Remember that the machine is still running and producing product.

Remember an out of control, like the previous change, can be either an improvement or deterioration. The control chart alarm system was never intended to imply that the product produced is bad. As a matter of fact this could be an improvement of the product! Since the alarm may not mean bad product, the machine will continue running until the strategic decision-maker has assessed the impact.

First the operator must identify the metrics for research and what the targets are for each metric.

The operator builds a research plan. Four metrics will be identified and checked. These are nut density, chocolate spray rate, nut drop rate, and conveyor speed. Next, the sequence for our checks is identified. First the operator checks the chocolate spray settings, second the nut drop rate, third the conveyor speed, and forth the nut density.

Shutting the machine down makes it difficult to identify the cause. Actually, this research is almost impossible with the machine shut down. Of our four metrics, only peanut density can be checked when the machine is shut down.

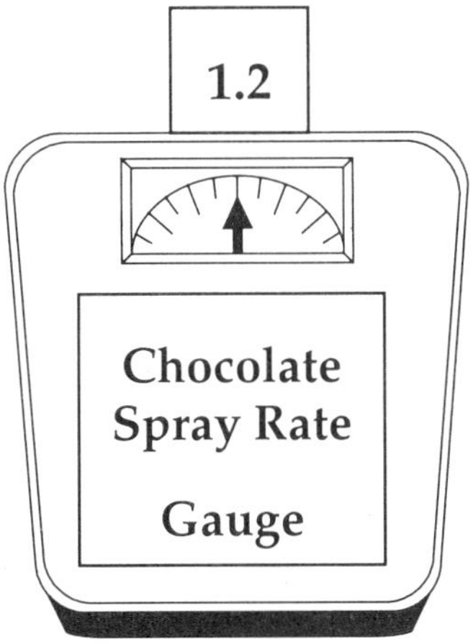

Our operator goes to the chocolate spray measurement to check its reading. If too much chocolate is dispensed, the clusters will weigh heavy. If too little chocolate is dispensed, the clusters will weigh light.

The target setting for the spray is 1.2 and the dial on the sprayer is set to 1.2. The actual amount of chocolate being dispensed is also the target amount of 1.2. From the operator's research this leads to the conclusion that the amount of chocolate on the clusters is not the cause of the weight having dropped.

Now we continue our search for the cause of the change. Notice we are only control charting our cluster weight and none of our four process metrics.

Having ruled out chocolate spray, our operator now moves to our second metric, nut drop. Our concern now is the amount of nuts being dropped. If the rate is too fast, the clusters will have more nuts than desired causing the weight to be too high. If the rate is too slow, the clusters will have fewer nuts than desired causing the weight to be too light.

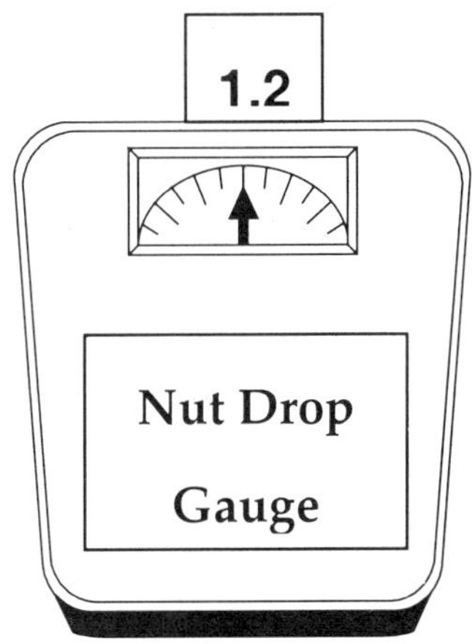

The target setting for the nut drop is 1.2. The dial on the nut dropper is set to 1.2 and the amount of nuts dropping measures 1.2. Now we have eliminated another metric from the potential causes of the lower cluster weight.

Our search continues with two possible causes being eliminated.

New measurements are taken and recorded on the control chart. The measurement of our product weight truly indicates a change is real.

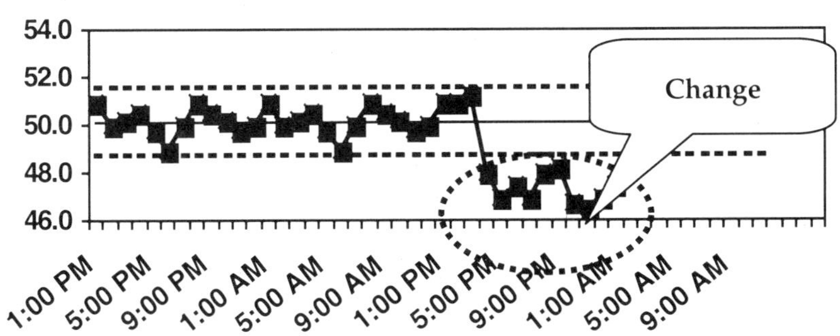

We must continue our search for the cause of the change.

The role of the strategist

While the tacticians search for the cause, the strategic decision-makers assess whether the change is improvement or deterioration. With more results that verify the change, our capability study becomes an accurate picture of this machine's current state of production. The pictorial capability study is shown below.

The capability study predicts that ten percent of our production on this machine is going to be below the lower specification limit. When ten clusters are produced, one of them will be too light to ship. Once this is known, the new question is which one of the ten is lighter than our reject criteria and which nine are shippable.

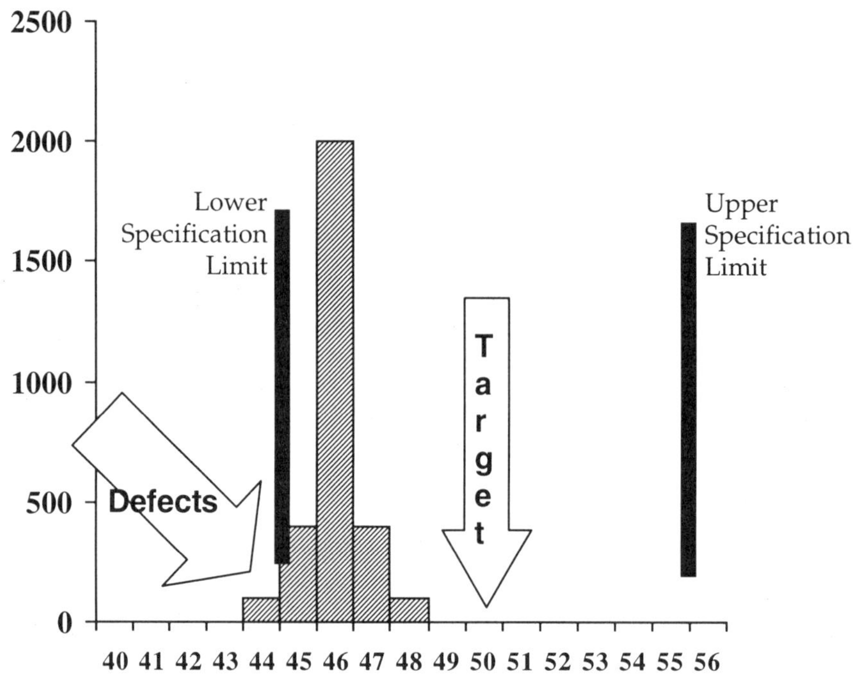

Notice, the total production picture shown above. All of the clusters are well removed from the target. As a matter of fact, none of the production is on the target.

Knowing that ten percent of everything we produce on this machine is too light to ship, we must now find the cause and figure out what to do with the clusters produced since 4:00 when the change started.

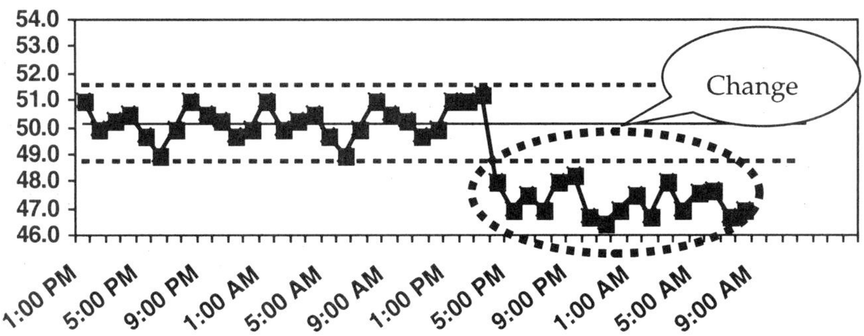

The role of the operator - still searching for the cause

Our operator is ready to look at the third metric, conveyor speed. If the conveyor carrying the cluster under the chocolate spray and the nut drop is too fast then too few nuts and too little chocolate are applied. The resulting fewer nuts and less chocolate will make the cluster weight light.

If the conveyor carrying the cluster is too slow then too many nuts and too much chocolate are applied. With this extra, the cluster weight will be too heavy.

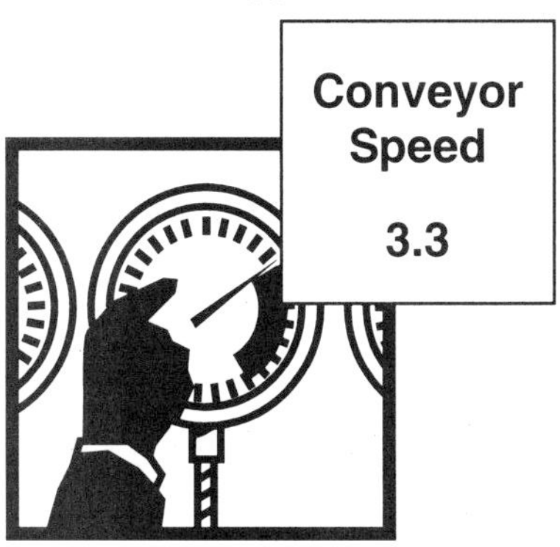

The target setting for conveyor speed is 3.1. The dial on the conveyor is set to 3.3 and testing the speed shows a measurement of 3.3. Thus, the dial is calibrated properly to the conveyor speed. With the conveyor going faster than expected, the nuts and chocolate being dispensed are not synchronized.

Now we have a reason for the weight shift. The operator adjusts the conveyor speed to the target setting of 3.1 and we will watch and measure the results.

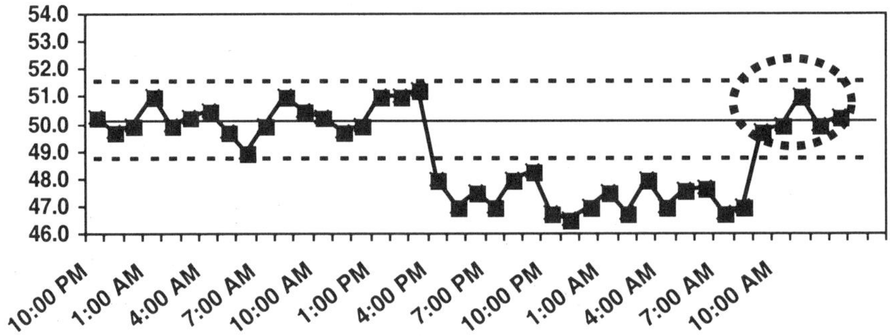

After the conveyor speed adjustment, the resulting measurements are shown above. The machine's weight metric now moves back to its original state and is returned to correct and consistent running parameters.

The final task is to deal with the product produced from 4:00 pm through 8:00 am. Of our 42,000 contaminated clusters 1,400 are too light to ship.

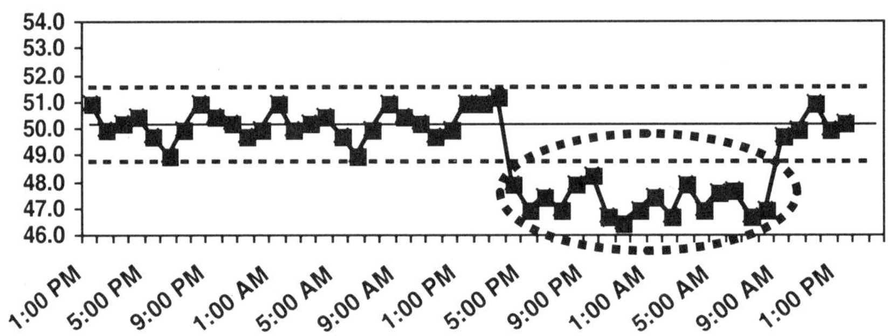

Our strategic decision-makers must establish a facility to quarantine this production. Then they must develop a method for light cluster removal, and they must staff this effort.

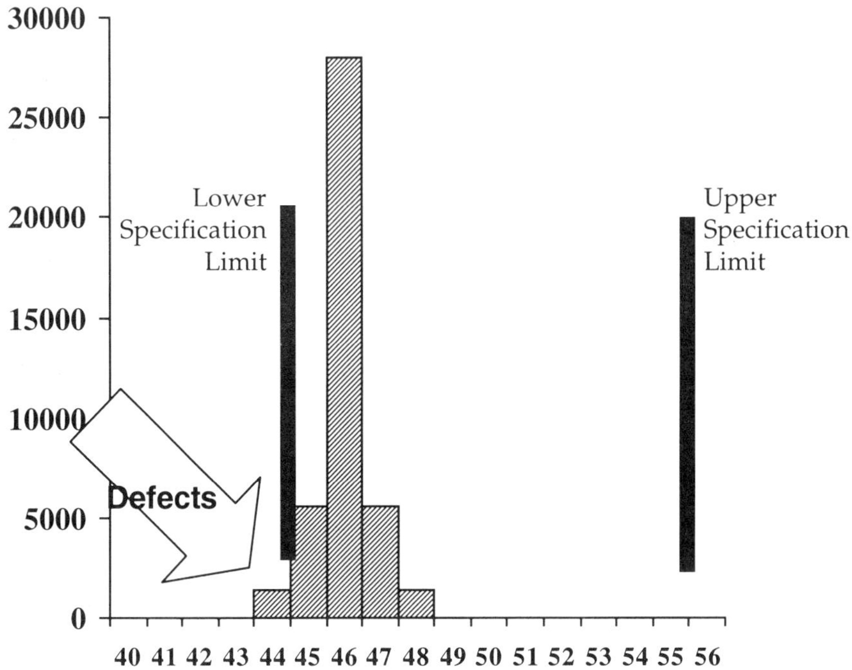

The strategic decision-makers designate a facility for damage containment. In this facility we test the weight of all 42,000 clusters. Those clusters weighing less than the specification limit (45 grams) are removed. This allows us to salvage the acceptable clusters.

As we mentioned earlier, the problem with this approach is the added cost, the delay time, and the possibility of some bad product getting past our test method.

In a later chapter we will discuss a better approach to running your machine. This more effective method is the true, correct, Statistical *Process* Control (SPC).

Chapter Seven:
Tools for Running your Machine

This chapter is designed to explain the use of the tools to optimally run your machine. The target audience for this book and this chapter in particular is the tactical workforce. The tactical workforce's mission is to optimally run the provided machine and this chapter details the tools necessary to accomplish that mission. This chapter is not intended to show how these tools are developed nor their underlying theory. Should you want to study these issues other books in the Walkabout® series cover those topics. Two of the books in the series that are of particular importance are:

- ***Optimize Your Operation:*** *Stories, Tools, and Lessons for Using the Principles of Process Management to Improve your Quality*

- ***SPC:*** *Practical Understanding of Capability by Implementing Statistical Process Control, third edition*

Optimize Your Operation explains the details and subtleties of the Principles of Process Management, the development and construction of the Walkabout® Dependency Diagram and Walkabout® Method, and the impact of Division of Labor on decision-making and goal setting. This book explains the method for achieving strategic correctness for each machine.

The book *SPC* deals with the latter two components of the base-camp: the consistent running of each machine, and an evaluation of the machine's capability. *SPC* gives a comprehensive explanation of the theory, practical uses, control chart construction, capability studies, and implementation; it was designed for the process engineer or quality professional.

Guiding concepts for an optimally run machine are Walkabout® Base-Camp, First Principle of Process Management (Product and Process), and Second Principle of Process Management (Division of Labor). Let's take each concept and explore them in detail. Then will discuss the tools that support these concepts.

WALKABOUT® BASE-CAMP

Starting in the front of this book we have built the case for running a business, or an improvement process, like an expedition ascending a mountain to its summit. Running our machine is also like that mountain climbing effort. A mountain climbing expedition always meets roadblocks. The expedition requires certain essential elements to overcome these obstacles and so does the improvement process. The mountain climbing expedition's essentials include the right equipment, access to the essentials of life (food, water, air), a stable launch point, and properly trained team members. These essentials are referred to as the base-camp for an expedition, and in our improvement process, a base-camp will also be required. The operational base-camp's components are correctness, consistency, and capability which we will refer to as the Walkabout® base-camp.

Correctness

We, the tactical workforce, are dependent on our strategic decision-makers. First, the strategic correctness must be established. A correct, functional organization with clearly defined roles must be determined to cover marketing and sales, product design, process design, manufacturing, and quality assurance. The product and process factorial families must be correctly calculated. A clearly defined product and a detailed method must be created. The strategic decision-makers must clearly communicate these correct methods. The strategist will provide a Walkabout® as schematic diagram of the method. On each metric a target will be provided.

We must brilliantly execute the plan, methods, and target settings. This brilliant execution should not be confused with perfection. We as human beings will always strive for perfection but the reality is that it may be outside our grasp. Correctness is the first step toward building the Walkabout® base-camp.

Consistency

The second step toward completing our base-camp means monitoring each metric for every machine. This will determine how consistent each product and process metric is. As we tacticians strive for perfection, this consistency provides a way of predicting what will be produced time after time. Each machine must have its own consistency established. This consistency is monitored through the use of mathematical tools called control charts. In earlier chapters we covered the posting and reading of these charts; now our attention is directed toward the use of these charts for running our provided machine.

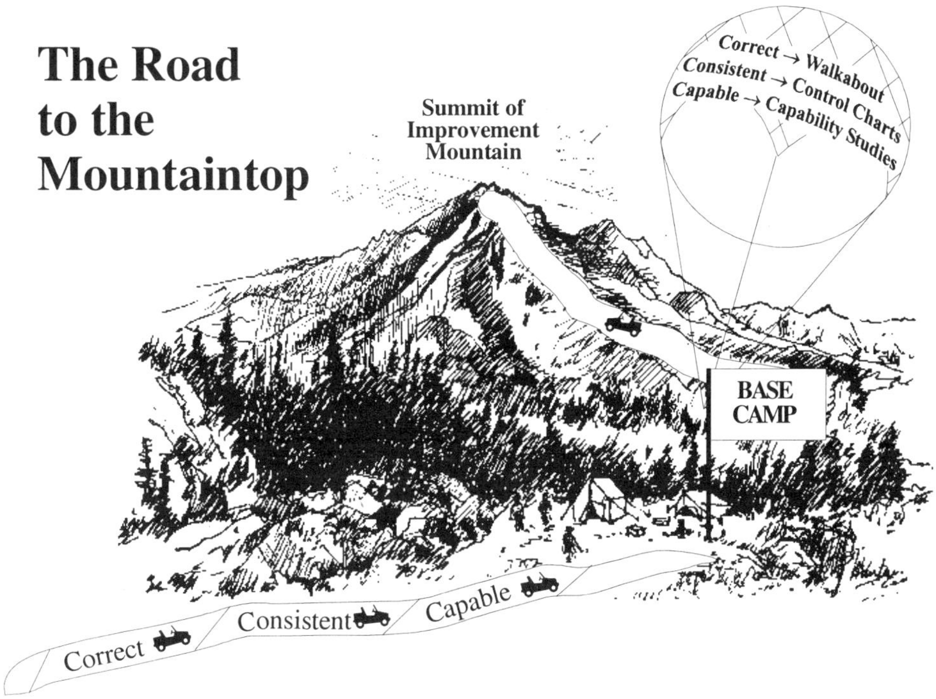

Capability

Finally, with the steps of correctly and consistently run machines or processes, the strategic decision-maker is positioned to assess how well the product we make meets the customer's expectation.

Take special note: Correctly and consistently running the provided machine are tactical issues. Capability is a strategic issue. Both the tactician and strategist must work together with brilliance to achieve the best performance from each machine.

FIRST PRINCIPLE OF PROCESS MANAGEMENT

To run a machine optimally, we must understand the First Principle of Process Management. Clearly understanding this principle is mandatory to everything we will do.

FIRST PRINCIPLE OF PROCESS MANAGEMENT

A fundamental understanding of BOTH the product and process is essential to improvement. Both the product and the process must be described and understood individually and separately. **The underlying component for improving the product is the process.**

The First Principle's final sentence is our key. This sentence reads, **"The underlying component for improving the product is the process."** Often all our efforts and energies are directed at analyzing the product. As an example eating a lot of biscuits does nothing to the baking effort. Having the ability to differentiate between good and bad biscuits does not make you a fine baker. A baker must focus his attention on the baking process. The temperature of the oven, how long the biscuits are in the oven, location of the baking tray, etc. are the key components for the baker. These are the measures of the process; for the tactician, they are significantly more important than eating the biscuits!

Obviously, the person running the machine must focus their attention on the process, and their knowledge of it must continue to increase. This expanded knowledge requires clear techniques that will be described in this section. Grandma told me the only way to learn how to make biscuits was to get into the kitchen and learn about the oven. Similarly, to improve our product, we MUST understand our process and this understanding must be converted into a process knowledge base that we can use to improve our product.

The Importance of Process Data: When we just monitor the product, we are aware of its deterioration or improvement only. We might be aware of the change but we do not have a reason for it. Product monitoring is important, but it is reactionary, tolerates waste with no understanding of why, and accepts improvement with no understanding of why.

Without understanding the why we cannot operate our machine to its maximum performance. For us to make Grandma's biscuits, we must understand why things occur the way they do. Product information gives no insight into why we made good biscuits or why we burnt the biscuits. Since we don't know why we are making good biscuits, we cannot repeat the activities that produced them. This loss of knowledge is tragic.

When we monitor the process and the product, the process parameters allow us to know the conditions that caused the change. Monitoring both the product and the process is essential to proactive decisions, improvements, elimination of waste, and understanding the "whys" of change.

Knowledge of "why" allows us to improve continuously. By adding knowledge about the activities required to make a biscuit (process), we begin to learn what is required to make good biscuits. The additional information about the different processes that were used for good biscuits and defective biscuits allows us to continue to repeat the improvements and to discontinue the defects.

In the perfect world of my grandmother and her biscuits, I never saw her check the biscuits when she removed them from the oven. She kept her focus on the process. Since she knew all the subtleties of baking, she would take corrective action as the oven was cooking the biscuits so that, in a proactive manner, she avoided ever making any burnt biscuits. She was a perfect chef; but unfortunately, most of us are not perfect chefs of our processes.

Since we live in a not-so-perfect world, and our knowledge of our process is not nearly so complete as my grandmother's was about her oven, we must continue to monitor both our biscuits and our oven with equal intensity. More precisely, we must monitor **both** process and product.

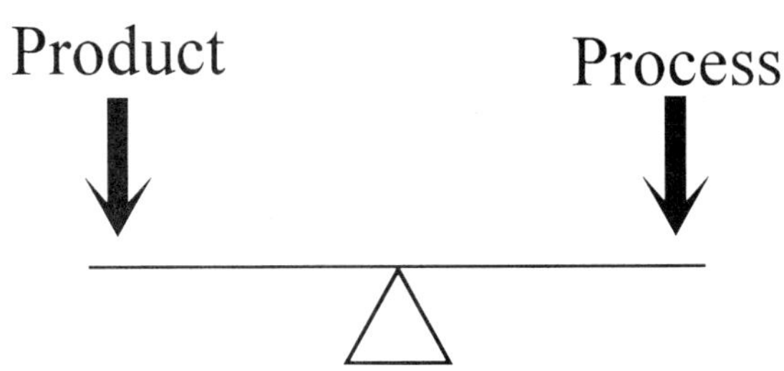

As we gather data and knowledge, we must accumulate equal amounts of knowledge about both the product and the process. This will allow us to build a knowledge base for both the product and the process.

Increased analysis of the product does not equate to increased knowledge of the process. Knowledge of the process and product are two separate but equal efforts and results. Both product and process knowledge is necessary for improvements.

JUST GOOD ENOUGH VERSUS ALL YOU CAN BE

When we are traveling on an airline, our objective is to arrive at a particular destination like Denver, Colorado. When we go to the store for milk, we want one gallon of milk. When a half-inch hole is drilled, we want the hole to be exactly a half-inch. In all cases we have something specific in mind. Better stated, we have a target in mind.

For clarity of purpose, our target must be clearly identified. The target must be explicitly defined, and a value for the target must be established. This target value is a strategic decision. Some refer to the true target as the nominal value. The **target**, simply defined, is the value that the customer expects should be perfect.

Knowledge of the target, or perfection, is another key to running our machine optimally. How can a plant, operation, or machine, ever achieve perfection if we do not know what perfect is? Worse yet, if we fail to communicate what perfect is, how can we ever expect to achieve perfection? Even if we know the target, will we be able to hit it consistently?

Unfortunately, we know that we will not be perfect. While we might sometimes hit the target, other times that will be impossible. How far away from the target can we move while still keeping our customers happy? To answer the question, we must determine tolerances.

The knowledge of what a customer will tolerate is important and defined by specifications and tolerances. This information is important but not for running our machine. Let's talk about this a bit and then come back and discuss what is important for us to run our machine.

Tolerances are what the customer will tolerate today or how much leeway from the target will be tolerated today. When the customer is describing his product, he communicates through tolerances and specifications. The customer has a specific item in mind. Specifications or tolerances are really the amount of *incompetence* that the customer will tolerate today.

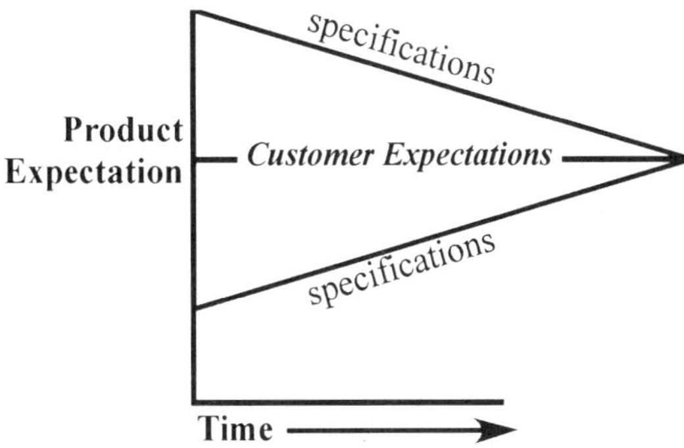

If we use the degree of incompetence to explain specifications, we should not be surprised to see the customer's expectation, or target fixed. We should also not be surprised to see the specifications continually changing. The tolerances are continually changing and getting tighter to drive toward the real expectations. As these tolerances tighten, the specifications get closer and closer to the target. This drawing demonstrates how specifications will change over time until ultimately the customer receives exactly what he wants.

To clarify the idea that tolerances and specifications will change, let's visit the grocery store to purchase a gallon of milk. Our milk buying specifications might prompt us to ask for two quarts to 1.5 gallons of something between buttermilk and water, rather than asking for one gallon of milk.

Properly established, the criteria should be a target of one gallon of milk. In the early 1900s, when milk was obtained in the barn from Bessy the Cow, the level of incompetence was somewhere between two quarts and six quarts which was all Bessy had.

What Bessy could and would deliver dictated the amount of incompetence that was acceptable in 1900. Over the years the amount of incompetence that is tolerated (the "tolerance") will decrease as the technology improves. When dairy plants began to bottle our milk the tolerance became smaller. Do not be misled that the customers will pay us more for these improvements; they will actually expect the cost and price to come down. In this example, the performance, cost, and delivery time to the customer will improve over time. In 1990 everything about dairy products is better than in 1900, and our customers' tolerance, or acceptance of incompetence, is much lower. As the product has gotten better, note also that the price has dropped.

In 1996, with milk in plastic and cardboard containers at the grocery store, our level of incompetence has shrunk. The target is still one gallon, but now the specifications are almost exactly one gallon. Over time, the incompetence index of the specifications will shrink. Also, this must be done without increasing the price.

We must keep moving toward the target, or the consequences will be severe. For the grocery store example, we must close in on the one gallon or risk having to always provide more than the customer's expected amount. Providing the extra, just to be sure we provided at least one gallon, will cause our cost to be too high.

The quality department identifies and removes defective parts using specification limits. The specification limits range can only tell us when a part is defective. This range cannot be used for operational purposes. Trying to use this specification range will lead us to poor productivity, inefficient machines, and poorer quality. Operations using this range can create the false sense that we are running the machine optimally or that we are doing well. This kind of tactical decision-making heads us straight to the *Just Good Enough Philosophy* of running the machine, department, or facility. To run a machine to its optimum, first let's talk about tactical decision-making and the impact of Division of Labor.

OPERATIONAL DECISION-MAKING

SECOND PRINCIPLE OF PROCESS MANAGEMENT

Division of Labor is the framework for all aspects of decision making. It must be clearly understood to separate the strategic and tactical decisions. **Operations makes the tactical decisions of running the facility.** Management makes the strategic decisions of assessing the facility's suitability for the job.

The Second Principle of Process Management Summarized: Division of Labor drives the decision-making process. It dictates who will be in charge of the different kinds of decisions (strategic or tactical). We are doomed to a state of Quality Hell if the managers of the future fail to fully grasp Division of Labor.

Tactics are defined as accomplishing the goal using available means. Control is a major tactical issue associated with operational duties. **Operations makes the tactical decisions of running the facility.** Capability is a major strategic issue associated with management, customer, and sales concerns.

Tactical execution is required for strategic decisions to be effective. If operations fail to correctly run the facility, no strategy will work. Our strategic decision-makers must count on our tactical decision-makers to do their job, for only then can the strategic decisions be effective. Many strategic options are available, but they will work only if the operation runs smoothly.

What not to do is just as important as what to do. Second Principle violations must be understood and avoided in order to effectively make decisions. The Good Trooper Award is given to the operator who tries to take the responsibility for strategic decisions. The harder an operator works with improper instructions (violating the Second Principle), the poorer the product quality. Quality will continue to degenerate.

When operations violates the Second Principle on an extremely capable process, we will make the machine appear less capable than it really is. Instead, we must run the machine to its own personality, as correctly and consistently as possible. We must not be lulled into accepting products that are only as good as the customer wants and allow the process to float inside the customer's specification.

We must guard against any violation of the Second Principle – Division of Labor. A good associate who is put in the situation of having to run a facility with improper directions that violate the Second Principle is doomed to disaster. The better and more conscientious the employee, the worse the situation will become because the person will go to any lengths to carry out those improper directions.

Running your machine is clearly tactical and the job of the tactical decision-maker. These tactical decisions are:

- Noting when a change in the process and product metrics has occurred.
- Alerting management when a change has occurred so that the strategic decision-maker can assess the impact of the change on the product or the process.
- Finding the cause of a change.
- Working with the strategic decision maker in deciding when to keep the machine running or stop the machine.
- Having the process knowledge to know when and what adjustments are required.

Peak efficiency is dependent on strategist providing a correct and capable facility. Our tactical work force must count on the strategic players to do their end of the bargain.

YOU CAN'T DO IT ALONE

Understanding who is responsible for each task is key to a successful organization. If operators try taking on every role, they risk not being successful in any role. Violating the Second Principle by compounding the responsibilities of management with those of operations, leads to chaotic decision-making. The strategic decision-maker can only ask operations to do what the facility is capable of doing. In this way, we can hold operations accountable for correctly and consistently running the facility. Management must never abdicate their responsibility to provide a capable facility.

Throughout this book we have discussed the importance of Division of Labor for decision-making, goal setting, and proper analysis. This Division of Labor clarifies the difference between strategic and tactical issues.

Let's discuss the climate in the Southern region of the United States. This will allow me to explain roles and the trust required for an effective team. This year in South Carolina, and pretty much the whole Southern region of the United States, a heat wave engulfed us. This is a typical kind of Southern summer. Our strategic decision-maker must select the proper tools to keep us cool.

I remember as a boy trying to stay cool in church. The local mortuary (strategist) supplied hand held fans to everyone in our church. In our heat wave using those fans never kept us cool. Maybe that is why the mortuary was the supplier; they wanted us to know who to call when we passed away from the heat. Motivation, violent hand motion, and rapid bursts cannot overcome the lack of a proper tool that will keep us cool. No matter how our operators worked with the fans the heat was unbearable. As a Southern fan operator we must accept the heat until a better facility is provided.

Later mechanical fans were introduced but the heat still won the battle. Only when air conditioning came along were the hot summer days manageable. Each of these provided cooling systems (hand fans, motor fans, or air conditioning) must be run to its maximum potential. As a fan operator in the South, we must correctly and consistently run the provided system. Then a fair assessment of each option's efficiency can be made. We expect our strategic decision-makers to continually invent, listen to suggestions, and try new options until ultimately an acceptable solution is devised. This acceptable solution is then built into a capable facility.

When strategic and tactical personnel both do their jobs, they form a team. Operations is in deep trouble when management does not do its decision-making job of providing the facility. If the facility (available means) is not capable, we are doomed to having dissatisfied customers and stockholders. If operations does not perform its role of running the facility (available means) correctly and consistently, then how can management assess the impact to the customer?

The proper support information must come together to build a finely honed partnership that allows for good, effective, clear decisions.

Operations must count on management to do their job of providing the proper facility for the job. Operations must trust that the strategic decision-maker will step up and take responsibility for the strategic job. Now the tacticians can focus all their energies on correctly and consistently running their machines.

WHAT DO YOU REALLY NEED TO RUN YOUR MACHINE?

First of all the objective is to like the army says, **"be all you can be."** In other words get everything possible out of the machine you are running. I am not talking about slower speeds or more expense but we are going to run the machine to its maximum potential. To achieve this we must have the proper tools. Using the wrong tool, or making do with what is available, makes doing your best impossible.

In my and your home we have many tools. Tools like cloth scissors, children's paper scissors, ¼" socket, ratchet, butcher knife, small flat head screw driver, steak knife, circular saw, chain saw, boat anchor can be found in just about every home. We do not expect to use every tool on every job. I would be stunned to see a situation that would require a chain saw or a boat anchor to be used in my kitchen.

Everybody should not use the same tools. Without the proper training and maturity many tools in your house are dangerous and will not give the desired results. When my son was a young boy, I would never have considered letting him use our chain saw. Even now that he is a man, training in the proper use of these tools is required.

In a factory we have two types of tools; one is physical and the other is intellectual. The physical tools are the machines, the tooling, the chemicals, etc. The intellectual tools are the knowledge of how to run these machines. Training for both tool types is required.

We will cover the intellectual tools needed to run a factory. First will be the selection of the proper tools. From our earlier discussion of the Second Principle, **"Operations makes the tactical decisions of running the facility."** The factory requires tools that are explicitly designed to run the available facility to its maximum potential.

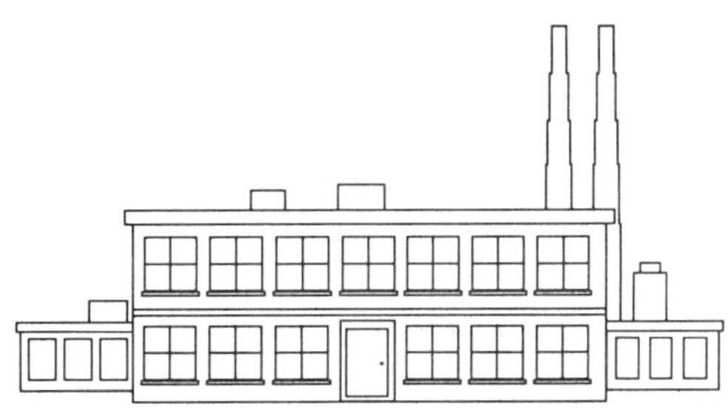

From the earlier discussion of the First Principle, **"The underlying component for improving the product is the process."** Now the operational workforce must focus on the process.

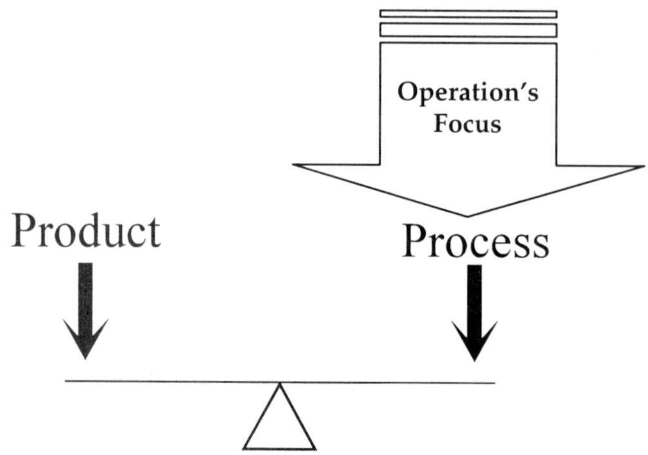

As we gather data and knowledge, we must accumulate equal amounts of knowledge about the product and the process. This will allow us to build a knowledge base for both.

Obviously, operations must focus all their attention on the process. These process activities are a proactive effort at avoiding problems and waste. My Grandma kept her attention on the oven and never had any burnt biscuits. Focus is hard to maintain. The tools that support the tactical decision-making of running your machine are needed to focus awareness on the process activities. These activities are setup, running, adjustment, and stopping.

The setup of the machine is the first step to correctly running the machine. The setup method must be clearly defined and communicated, and must be strictly adhered to by operations. Once the setup is complete, operations must run the machine following the correct method.

As the machine runs, tools for detecting a change in the machine's characteristics in a timely manner are essential. Once a change has been detected, the tactician must know what adjustments are feasible. Simultaneously, the operator must alert management to real changes so that these strategic decision-makers can assess the impact of the change. Finally, tacticians must appreciate that stopping their machine is a joint decision made with the strategist.

Let's now work our way through those essential tools required to support the operator in getting the most from their machine. These tools focus on the process, making sure that it is correctly and consistently run.

Operational tools

The operational tools for running a machine must focus on correctly and consistently running the process. The tools that support this mission are called Walkabouts® and control charts. The Walkabout® deals with correctness, and the control chart deals with consistency.

The Walkabout® communicates the correct equipment and tools for running the machine. Next the Walkabout® defines the method or sequence of events for running the machine. For each metric on the Walkabout® the product and process targets are defined.

Control charts are established for each machine and all of its metrics. The control chart allows the operator to see how consistently his machine can hit the targets. With this base line chart any deviation in consistency is immediately detected.

Running good product is dependent on the strategist providing a correct and capable facility. Our tactical work force must count on the strategic players to do their end of the bargain.

Let's revisit my childhood candy fantasy to explain the operational tools required to run a facility.

To assure our customers of quality candy, we have an unconditional guarantee that covers taste, freshness, ingredients, and weight. If there are any deviations, the customer will receive an immediate refund. Our purpose is to assure that no customer receives a candy that is too light. The guarantee states that no cluster that weighs less than 45 grams will be shipped. We must also protect the investment of the stockholders, for if any candies are manufactured that are too heavy, then we are giving away product and raw material, and this generates no compensation for the company because we are not charging a higher price for the extra ingredients. To protect our shareholders, we have set a specification of no cluster being manufactured that weighs more than 55 grams. Our specifications, therefore, protect both of our customers (shareholder and end consumer).

Operations: Before we proceed, I must explain the operations of the facility. Because the facility is state of the art, the engineers have detailed instructions for its correct functioning. These detailed instructions are spelled out in Walkabouts® provided by the engineers. When the facility is correctly and consistently run, 50-gram clusters (or as close to 50 grams as possible) will be produced. The person in charge of operations will be responsible for the correct and consistent running of the facility.

WALKABOUT® AS AN OPERATIONAL TOOL

The Walkabout® is our tool for communicating correct methods. Step one and two define the product and the process.

WALKABOUT™ Dependency Diagram

Area

Date/Version:

1 Product Definition

✓ Use clear succinct paragraphs
✓ Do not use "buzz words" or acronyms
✓ Use nouns

Round Chocolate peanut clusters

2 Process Definition

1. **Marshmallow Base Building**
2. **Caramel Drop**
3. **Peanut Sprinkling**
4. **Chocolate lay down**

ABBOTT ASSOCIATES INCORPORATED

150 Executive Center Drive, B37 · Greenville, South Carolina 29615 · Tel (864) 297-9598 · Fax (864) 297-8624

Call or write to purchase forms.

Our product is marshmallow covered with caramel, sprinkled with peanuts and covered with chocolate. The four activities for making chocolate clusters are shown on the Walkabout® process definition section shown to the left.

Next, we should describe each activity in more detail and the sequence of each activity. We must capture this information for our chocolate candy business and transfer that knowledge to the Walkabout®.

WALKABOUT® DEPENDENCY DIAGRAM

Our chocolate cluster process is very simple. We have a conveyor that travels past four stations. The Walkabout® below shows the four activities and their sequence. This simple schematic shows in pictures the sequence of events. The marshmallow base is built at the first station, and then passes down the conveyor to have caramel dropped on top. Next traveling down the conveyor, the cluster passes under a steady stream of falling peanuts. Any peanuts not applied to the cluster fall through the conveyor for reuse. Finally the cluster passes under a continuous stream of liquid chocolate.

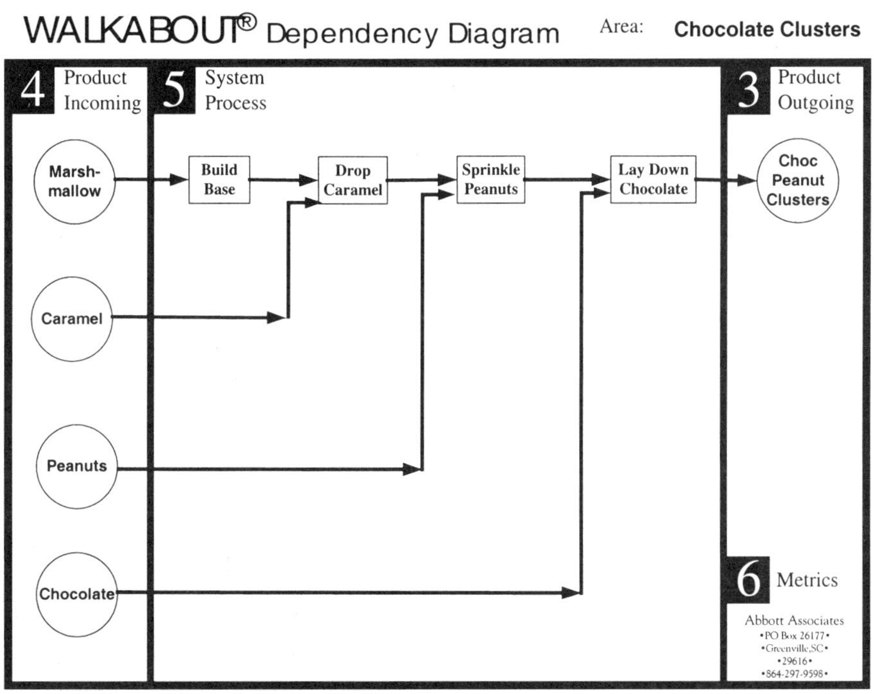

At each step of the cluster making effort, both cost and value are added. At any step our cluster could be destroyed with bad incoming products or poor process performance. Metrics will be the key to maintaining our product and process quality.

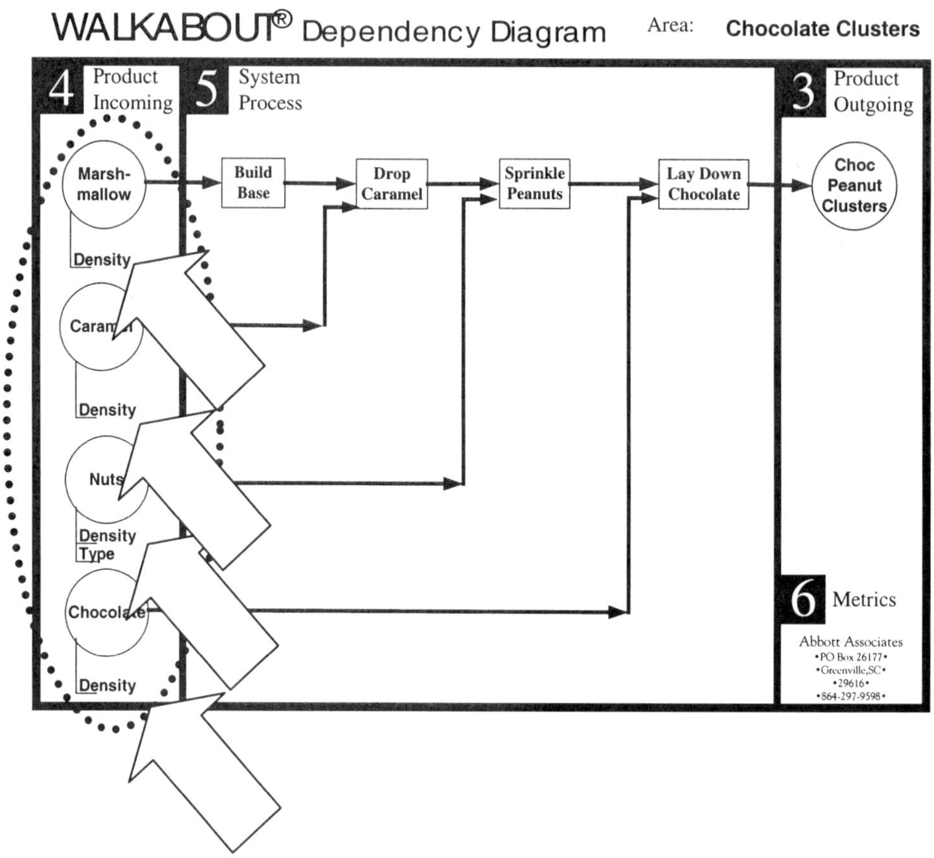

The four incoming products (marshmallow, caramel, nuts, and chocolate) must each have metrics. For our example only two metrics are used — type of nut and density. In reality, many more metrics will be in place to help us make our chocolate clusters and optimally run our candy machine.

More metrics are added for the processes. Since the incoming product comes from another machine, the process metrics are the prime concerns of the operator. Since the operator can't do it all, we must trust that we receive good ingredients. As Ronald Reagan said in his negotiations with the Soviet Union, "Trust but Verify." Put simply, we will trust that we are receiving good incoming product but we still check these incoming products because we do not want to waste the added value of any of the four steps. Should we detect a change in any incoming product metric we will immediately alert the strategic decision-makers for an assessment.

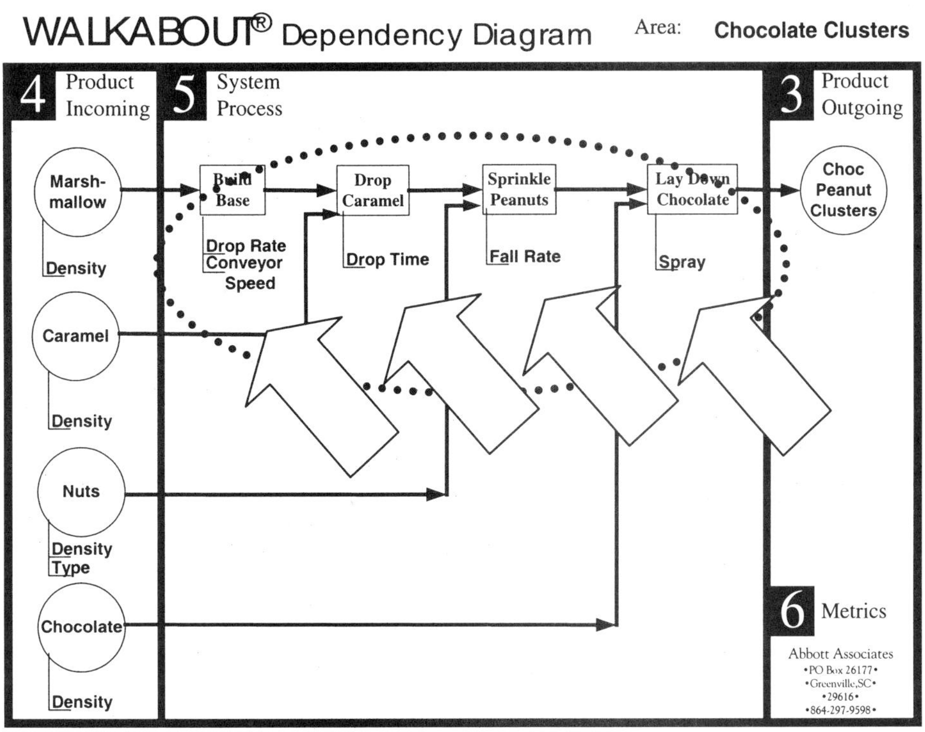

The metrics of the four processes are the operator's prime concern. For the base building process, marshmallow drop rate and conveyor speed are the two metrics involved. For the caramel drop process, caramel drop rate is the one metric involved. Since the marshmallow base builder and the caramel drop use the same drive mechanism, both processes flow at the same rate. This ensures that when the first process (base building) is correct, the second (caramel drop) will also be correct.

The rates for nut sprinkle and chocolate spray must be metrics. These rates must be synchronized with the conveyor speed. If the metrics are out of synch, then the final cluster will either be light or heavy on peanuts and chocolate.

Our final set of metrics is now added. These metrics are the outgoing product metrics. If the strategic decision-maker and tactical decision-maker have done their job, no measurement would be required. But recalling our *trust but verify* motto, we will add metrics for quality assurance.

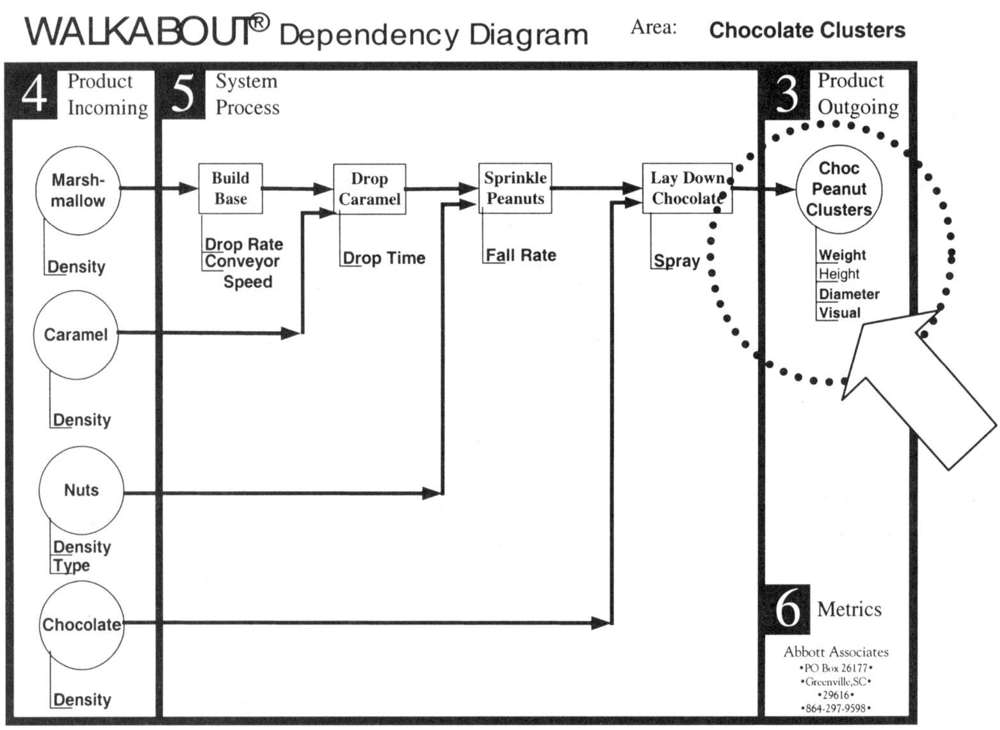

The chocolate cluster metrics are weight, height, diameter, and visuals (look, taste, and smell). We will monitor these for change, but any change in the product can only adjust any future production. Thus, the reason to focus our attention on the process is to avoid problems before they happen.

WHERE TO USE TARGETS

Several issues must be clear to truly understand how to run our machine. These issues will allow us to separate what the customer wants from our abilities to achieve their desires. Our desire to make the customer happy must not influence the tactical objective of optimally running the machine.

For clarity of purpose, our target must be clearly identified. The strategic product designer (strategic decision-maker) explicitly defines the product target. This target value is a strategic decision. Some refer to the true target as the nominal value. The target, simply defined, is the value that the customer accepts as perfect. Next our strategic process designer must engineer process targets that will achieve our product targets.

Knowledge of the target is key to manufacturing. How can a plant, operation, or service ever achieve perfection if we do not know what perfect is? Worse yet, if we fail to communicate what perfect is, how can we ever expect to achieve perfection?

The strategic decision makers must define the target. At no time should the tactician attempt to independently change this criterion. Any attempts by the tactician to take on the role of the strategist will lead to disaster.

The target is used to communicate our requirements. They are established specifically for operational setup. Once the machine is set up to the target value, we continue to run the machine to this target. Since each machine is unique, each machine then must have its own set of target values. These values must be continually evaluated, updated, and optimized to assure that this machine is running to maximum capability. These refinements are the responsibility of the strategic decision-maker as part of their role to provide a capable facility.

WALKABOUT® TARGETS

For each metric, our strategic product designer and process engineer have designed targets. These targets are now placed by each metric.

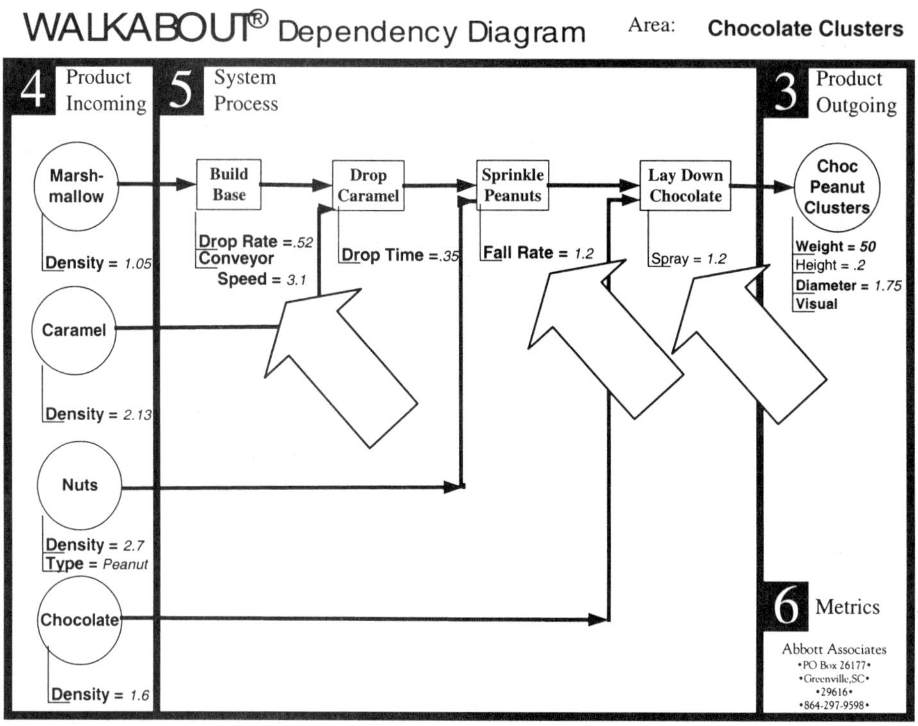

Each metric and its target are important. Earlier we said that both the nut sprinkle and chocolate spray rates must be metrics. These rates must be synchronized with the conveyor speed. If the metrics are out of synch, then the final cluster will either be light or heavy on peanuts and chocolate. Thus, for the operator of this machine, it is critical to monitor these metrics to assure good product.

Don't panic! I know you are not perfect and maybe can't hit the target every time, we will deal with those issues in just a few pages.

WHERE TO USE CONTROL CHARTS

A control chart is the tactical tool to monitor the consistency around our target. For every metric we must have a control chart monitoring our actual results. We will monitor all three types of metrics: incoming product, process, and outgoing product.

The control chart gives us a tactical decision-making tool to determine when any movement in our metric measurement is truly a change.

CONTROL CHARTS

First we set up our machine to the target. In this example our metric is the end product metric of weight, with a target of 50 grams. We run the machine to this target, gathering and posting the readings.

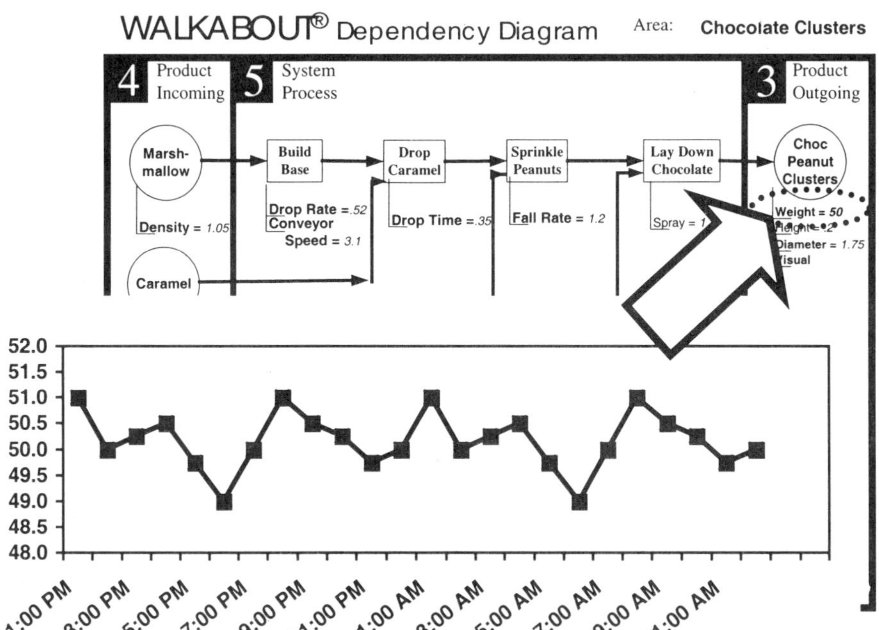

Next, based on some math done by our strategic decision-makers, control limits are calculated from the actual results of running the machine.

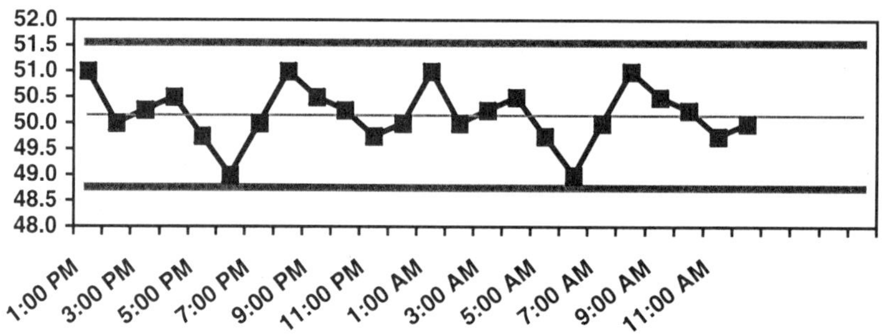

When the metric measurements on the control chart do not violate any of the SPC rules covered in Chapter Four, we can declare our machine to be in a state of statistical control. Using the Walkabout® to establish that we are running to the target, we have now passed the first two milestones in our journey to the Walkabout® base camp. Our strategic decision-makers will tackle the third milestone, "assessing capability."

The tactician has kept their end of the bargain. When the machine is running correctly and consistently, we are ready for the handoff to the strategist. A true team is built on trust. But we not only trust, we verify. The strategist must now take our measurements and assess the capability of the machine.

One term must be defined before the capability study can start. The term is called the specification limits. The specification limit is the spot where the customer judges the product is totally unacceptable.

The strategist does the capability study by superimposing the actual metric measurements onto the specification limits. The chart below is a capability study. The bar reflects the actual measurements showing the number of occurrences by group. The two lines are the specification limits.

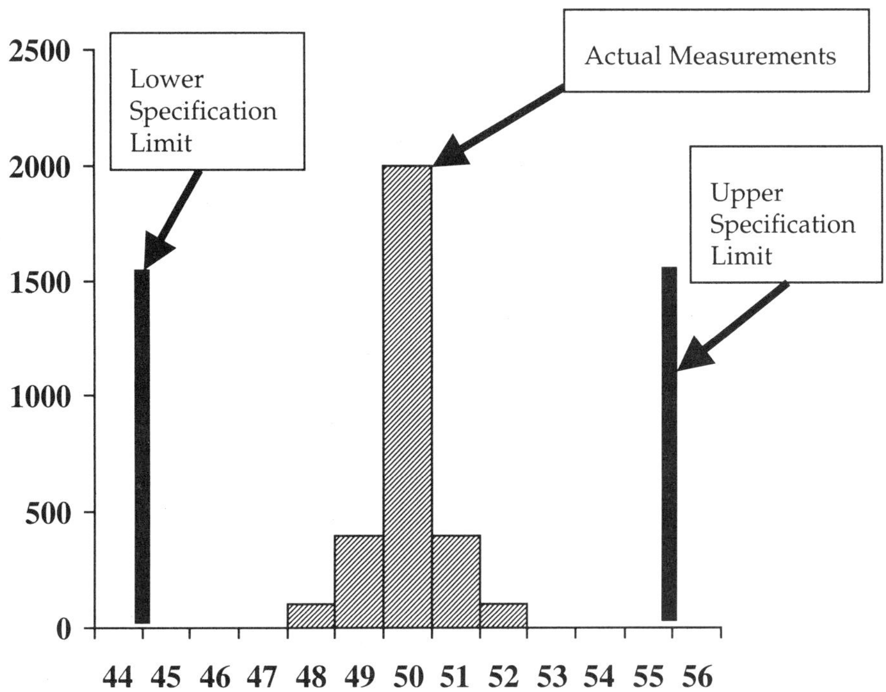

With the machine from our study, the measurements are shown with upper and lower specification limits. When the strategist has done the capability an expected zero percent defect rate is determined. As long as the process stay correct and consistent with no changes then we should continue to never see any defects.

Since capability is not a tactical issue, we will not get into the details of how this study is computed. Should you desire to find out the details of a capability study, the book ***SPC,*** *third edition* is available at your local or on-line bookstore.

In Chapter Five we covered the use of control charts, but let's do a quick review. When a change is detected, the operator must do two things. First, a study to determine the cause of the change must be initiated. Second, the operator must alert the strategic decision-maker to the change.

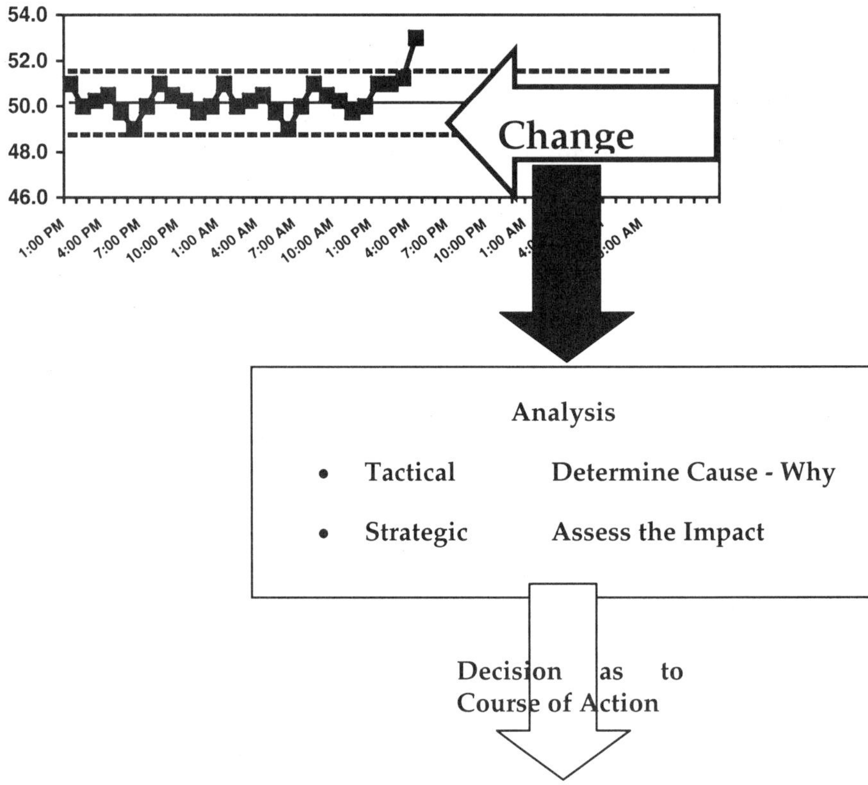

Determining why there was a change is a tactical responsibility. Assessing the impact of the change is a strategic responsibility. The answer to these two questions drives the response to the change. Change must not be confused with defective product. Change can be either an improving product or a defective product.

MACHINE UNIQUENESS

All machines are different, which necessitates a unique control chart for each machine. This uniqueness makes replication of the control limits from one machine to another impossible. The differences in machines can be because they were purchased at different times, purchased from different vendors, have differing maintenance profiles, etc.

The two charts below are for the same metric but from two different machines.

The first chart is for Machine A, a new machine. This machine produces results that are relatively close to the target.

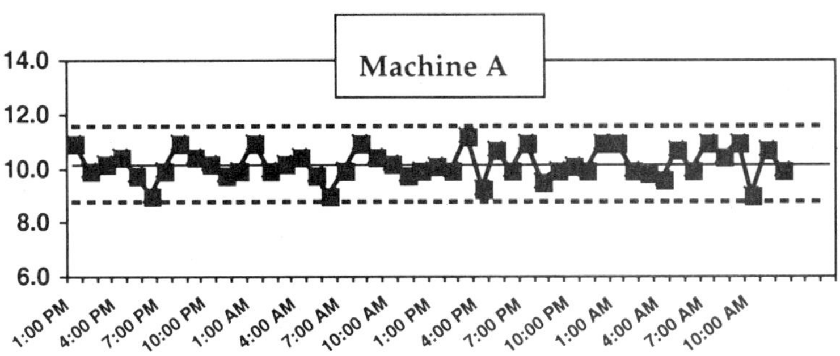

The second machine, named B, is an old machine. Compared to Machine A the measurements are much farther from the target.

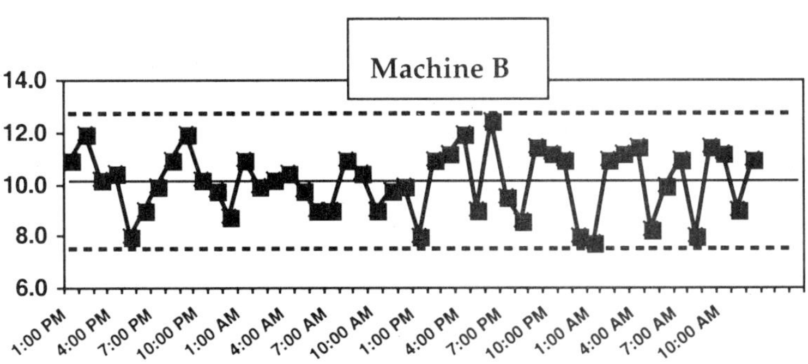

The control chart provides the medium for us to compensate for the uniqueness of each machine.

TACTICAL TOOLS IN REVIEW

The Walkabout® and control chart jointly form the prime tactical tools. The Walkabout® defines the correct method and metric targets. The target is used for setup and running the machine with the control chart.

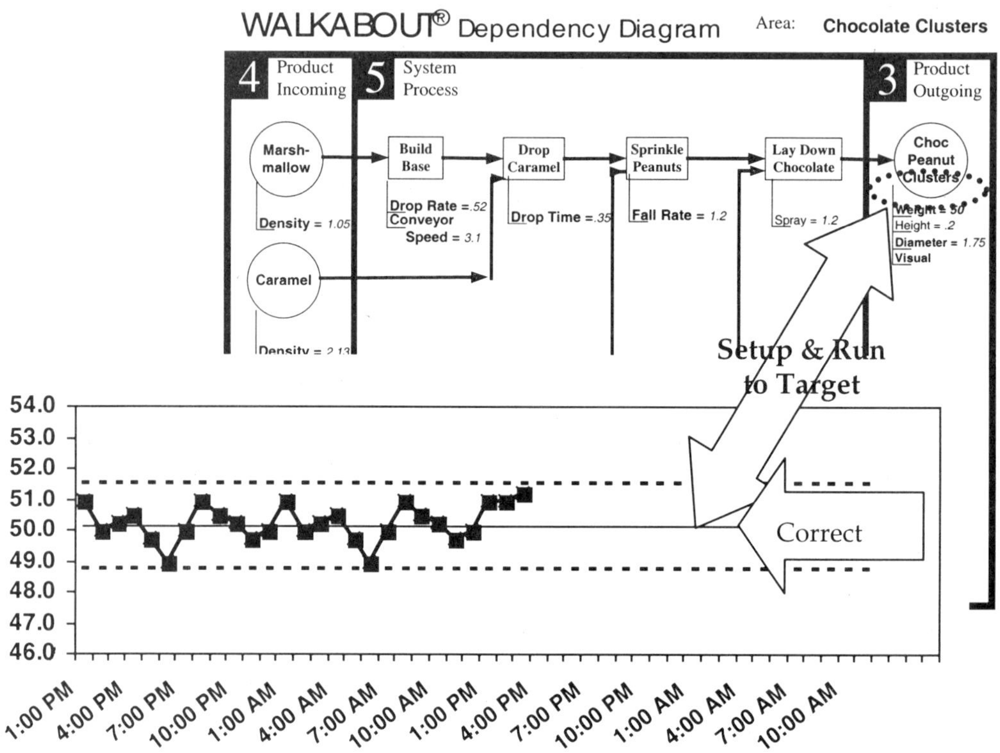

Execution of the strategically correct plan is key to running an effective machine. The control chart provides the guidance for noting changes in the personality of this machine's metric.

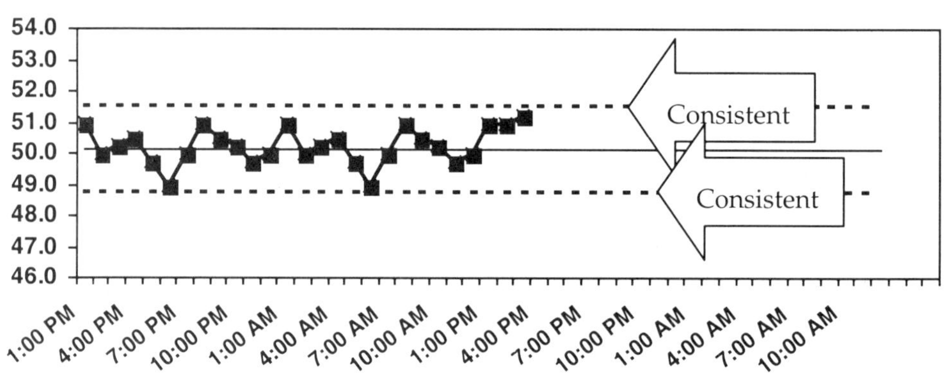

DEALING WITH MANY PRODUCTS

The tactical tools perform a perfect job for running a machine. As a facility adds more products into the mix, the volume of information to run your machine becomes massive. If we had to have separate Walkabouts®, targets, control limits, and control charts for each product and machine, the paper work would overload us.

Rather than simply having one candy product, let's add a new product to our candy line. We will add Cluster Supreme. The good news is the market potential is great, but the bad news is that each new product must have its own targets and Walkabout®. The Walkabout® below shows the changes that are required to the Chocolate Cluster's Walkabout® and targets to provide the information for our facility to make the Cluster Supremes. The Supremes are made with pecans and require a new nut sprinkle rate.

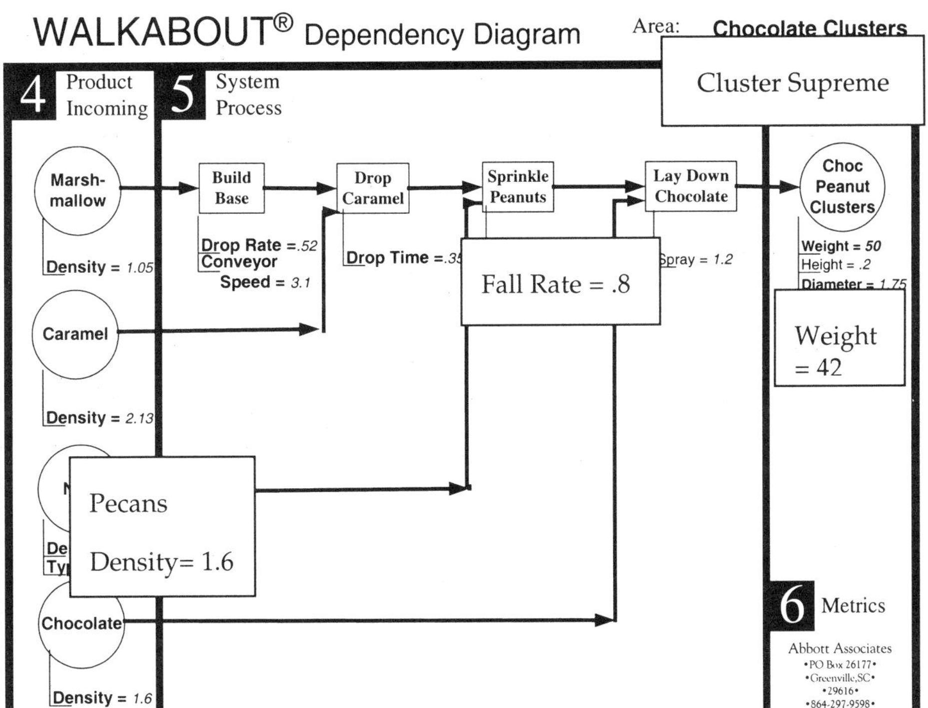

The new Cluster Supremes are produced and the end product with its weight of 42 sells very well. Our success pushes us to introduce more products.

More new products are added to our candy line. We add Cluster Brazil and Cluster Almonds. The market again is great, but the bad news is that each new product must have its own targets and Walkabout®.

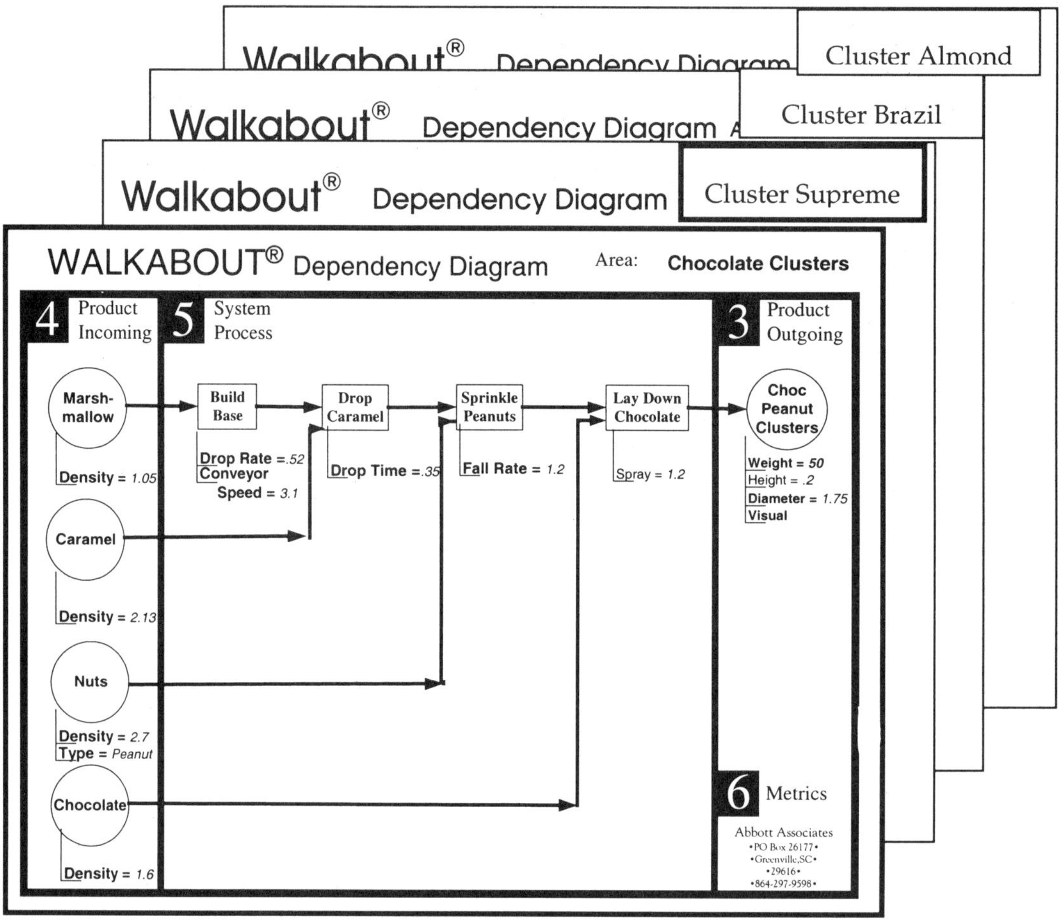

The Walkabouts® and targets continue to grow with each new product.

Imagine a metal machining plant that makes motors; the motor combination could be in the thousands and the parts for each motor could be in the millions. The combinations and number of parts are huge. Every part number would require a set of all these tactical tools. The good news is this is really not necessary.

There is an old saying, "It's easy to make things hard, but it takes really hard work to make things simple." An approach exists for grouping like items, products, parts, and processes. It is called factoring. The result of this sophisticated model is called a factor table.

WALKABOUT® FACTOR TABLES

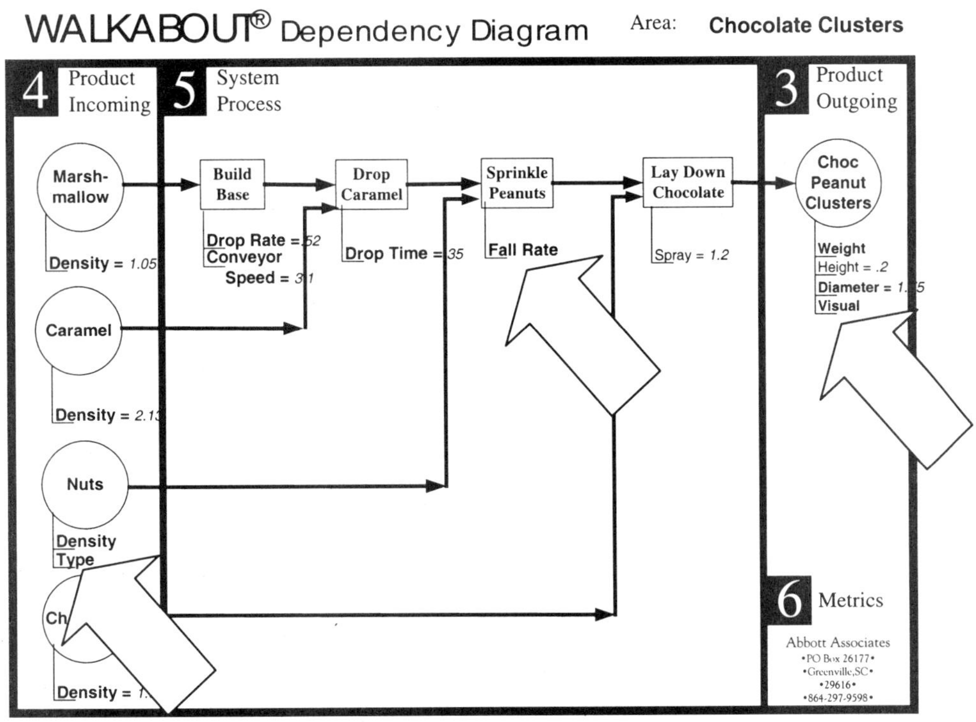

We have broadened the product line for our candy plant. The basic product is exactly the same – only different nuts are used. The processes are exactly the same as our original chocolate cluster. By adding different nuts, three metric targets must change based on the kind of nut involved. The three metrics are nut type, nut sprinkle fall rate, and product weight.

The table below is the result of the factoring process. Rather than having a Walkabout® for every product or part, one Walkabout® is constructed with a factor table in the corner of the Walkabout® page.

Product	Nut Type	Nut Density	Nut Sprinkle Fall Rate	Product Weight
Cluster	Peanut	2.7	1.2	50
Cluster Supreme	Pecans	3.3	.8	42
Cluster Brazil	Brazil	3.7	.6	40
Cluster Almond	Almonds	3.5	.9	44

The Walkabout® below includes the factor table.

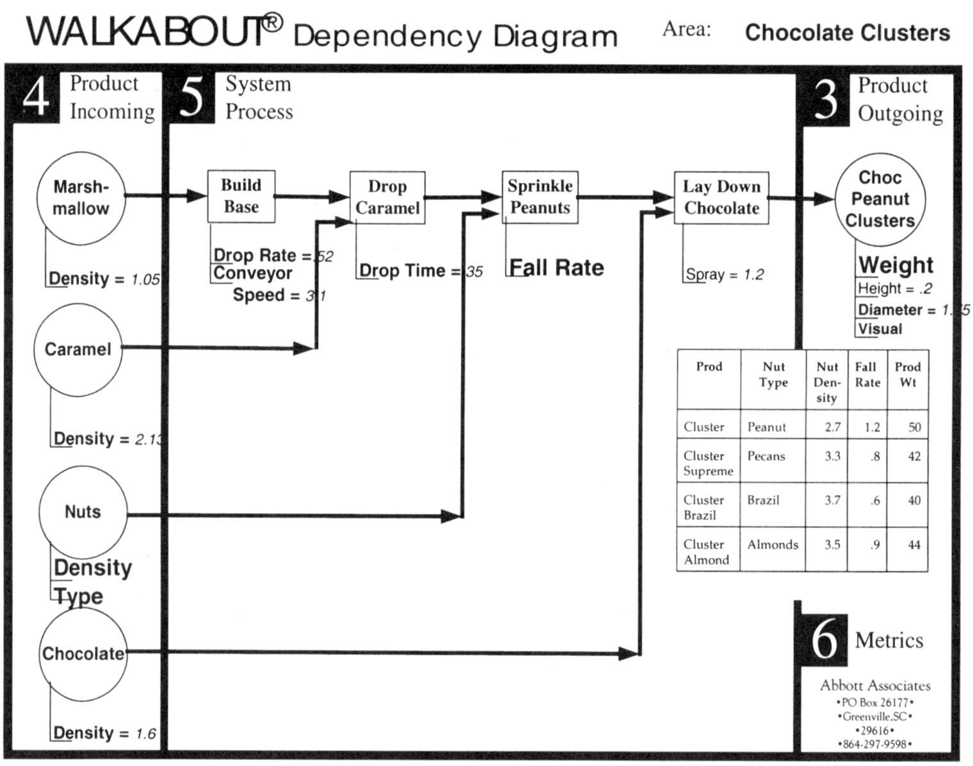

CHAPTER SUMMARY

An operation requires the following tools to achieve an optimum process:

- Strategically defined correct method communicated using Walkabouts®
- Targets for all metric settings
- Control charts for all metrics

Our prime focus as tacticians must be on the process. The Walkabout® helps us keep our eye on the ball and maintain the focus required for maximum performance.

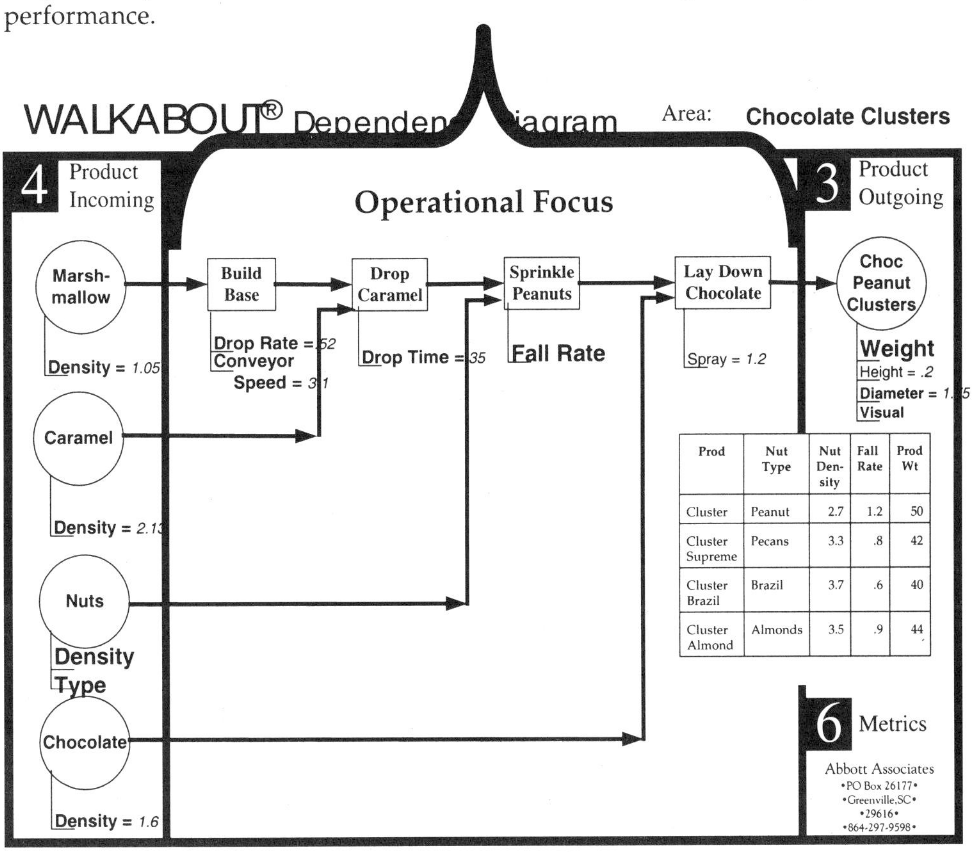

Prod	Nut Type	Nut Density	Fall Rate	Prod Wt
Cluster	Peanut	2.7	1.2	50
Cluster Supreme	Pecans	3.3	.8	42
Cluster Brazil	Brazil	3.7	.6	40
Cluster Almond	Almonds	3.5	.9	44

The terms below are key for the tactician.

Term	Use	Definition
Target	Operational Setup	Perfect!
Walkabout®	Running machine	A tool to communicate the method for running a machine and its metric targets.
Control Limit	Determining change	A machine's calculated limits for detecting change.
Control Chart	Detect change	A chronological sequence chart to detect change on a particular machine.

Next, we will use examples from our candy plant to show how to use these tactical tools.

Chapter Eight:

Running Your Machine with SPC

USING THE WALKABOUT® BASE-CAMP

We are ready to see how an expedition can ascend a mountain to its summit by using the right tools. Our mountain climbing effort is our machine. We introduced the essential tools in the prior chapter and we are ready for our first look at how to use these tools on our candy-making machine. The operational components are correctness, consistency, and capability, which form the Walkabout® Base-camp. The tactical tool for correctness will be the Walkabout®. The tool for consistency will be the control chart. We will build a series of chocolate candy plant stories that will allow us to see the tools at work.

Since communication is key, we will repeat the tactical terms.

Term	Use	Definition
Target	Operational Setup	Perfect!
Walkabout®	Running machine	A tool to communicate the method for running a machine and its metric targets.
Control Limit	Determining change	A machine's calculated limits for detecting change.
Control Chart	Detect change	A chronological sequence chart to detect change on a particular machine.

SETTING THE STAGE

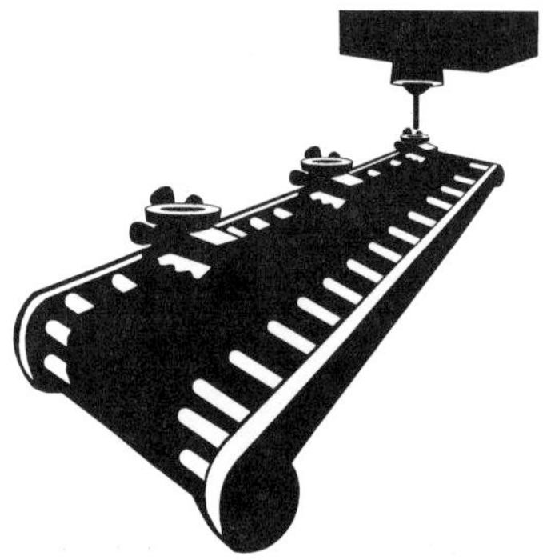

The conveyor carries our buildup of chocolate clusters past four stations. The Walkabout® below shows the four activities and their sequence. This simple schematic pictures the sequence of events. The marshmallow base is built at the first station, then passes down the conveyor to have caramel dropped on top. Next, traveling down the conveyor, the cluster passes under a steady stream of falling peanuts. Any peanuts not applied to the cluster fall through the conveyor for reuse. Finally the cluster passes under a continuous stream of liquid chocolate.

The operator's focus should be on the process; our machine's Walkabout® helps us keep that focus which is required for maximum performance.

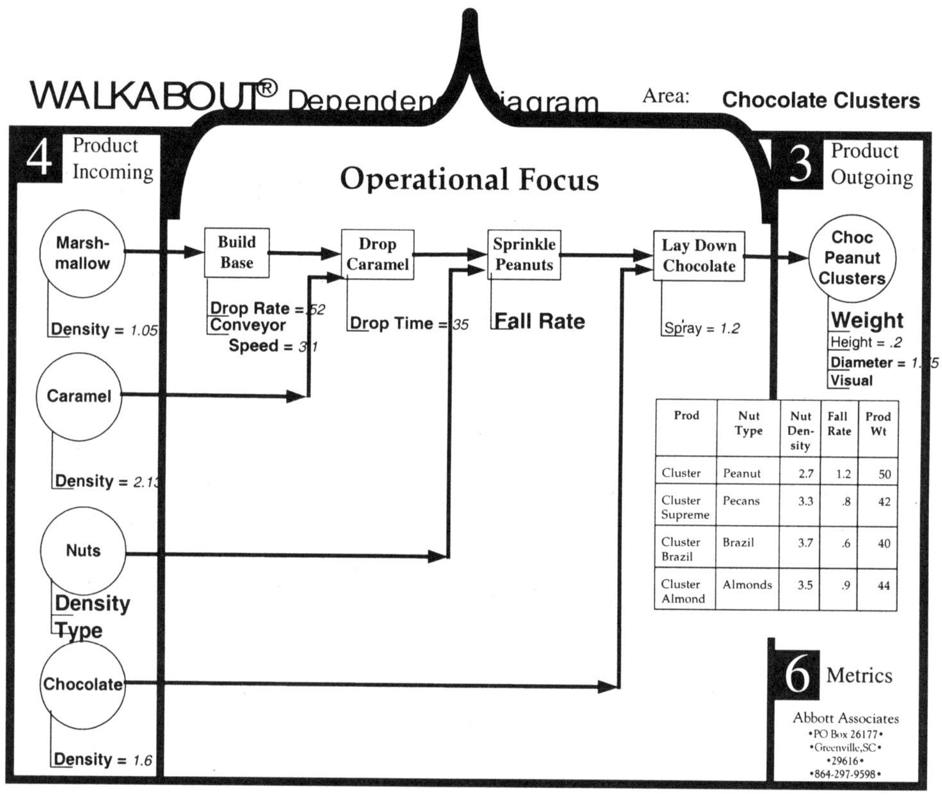

Our process has been designed to make four kinds of chocolate clusters: peanut, pecan, brazil, and almond. To make our stories more direct, pointed, and clear we are going to reduce the number of combinations that we will have to work with.

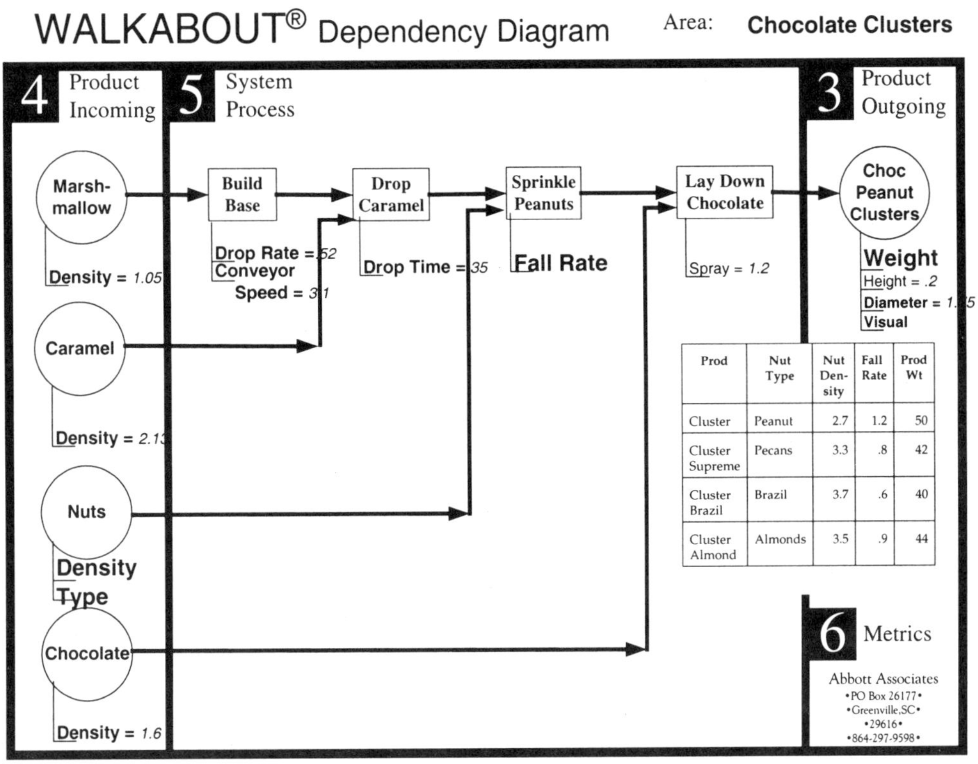

Prod	Nut Type	Nut Density	Fall Rate	Prod Wt
Cluster	Peanut	2.7	1.2	50
Cluster Supreme	Pecans	3.3	.8	42
Cluster Brazil	Brazil	3.7	.6	40
Cluster Almond	Almonds	3.5	.9	44

We are going to narrow the number of products for this chapter to the chocolate peanut clusters.

By reducing our products to one, we can revert to our simple chocolate cluster Walkabout®. Every in-coming product, process, and out-going product metric should be tracked, measured, monitored, and analyzed.

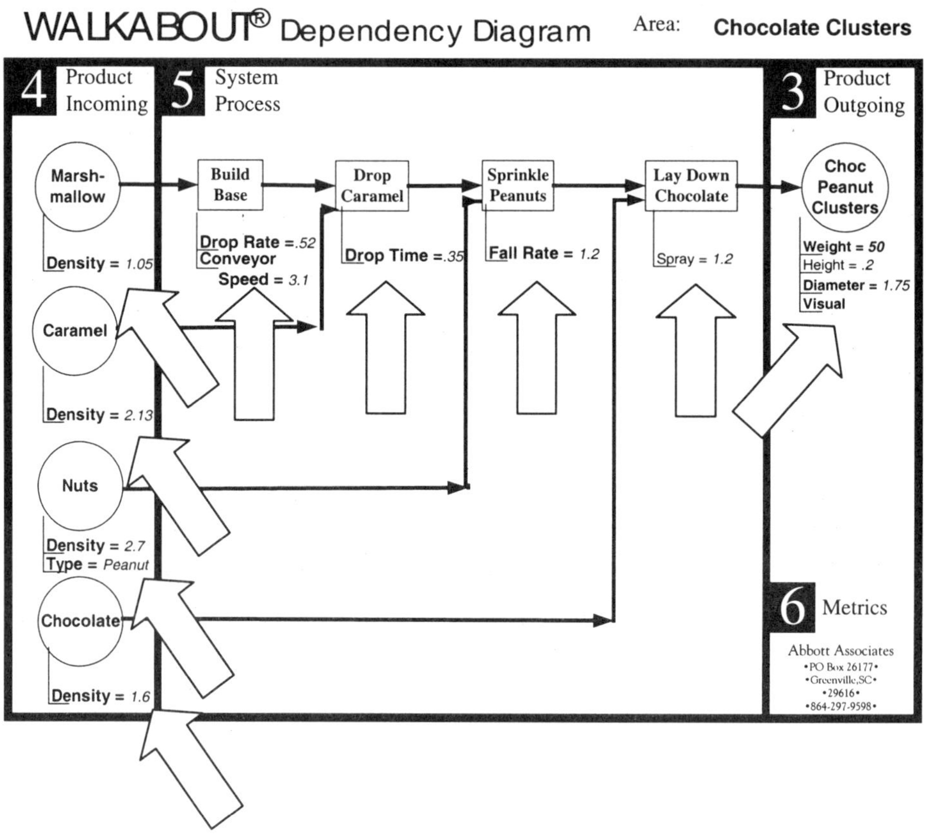

As you can see, the amount of information is large. If we were truly running a chocolate cluster machine, we would track and post all of the metrics on a control chart. For our example, we will reduce the number of metrics.

Earlier we explained that the first two stations, marshmallow and caramel base building are connected with a direct drive making improper base construction almost impossible. The nut drop and chocolate spray are a different matter all together. The nut drop is a continuous flow of nuts and must be synchronized with the conveyor speed. The chocolate spray is also continuous. The chocolate spray must also synch up with the conveyor speed.

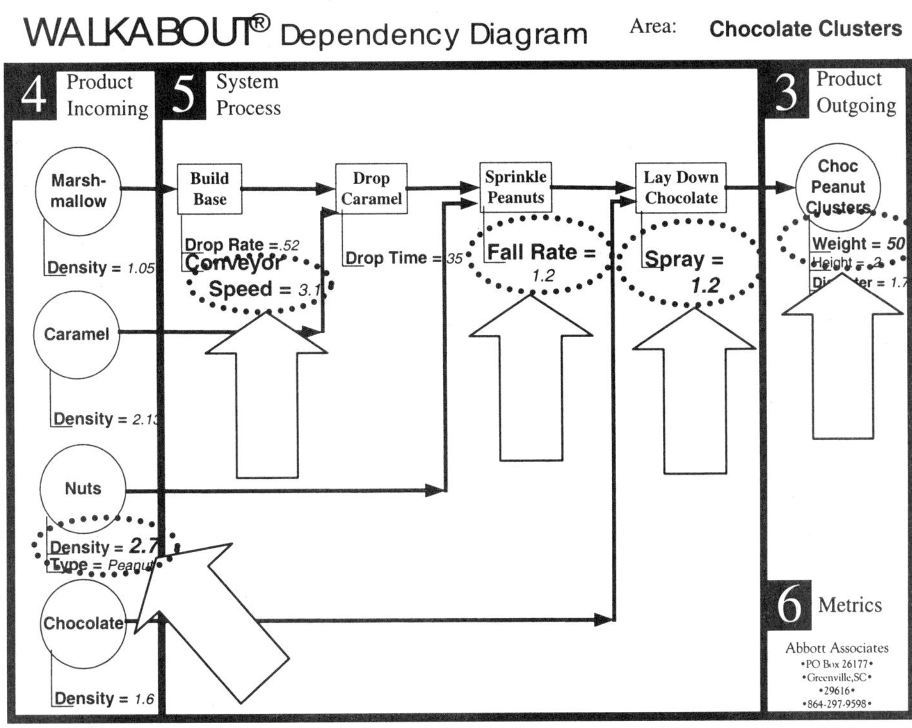

For our analysis, we will focus on five of the possible metrics. These metrics are: outgoing product's weight, incoming product's nut density, process conveyor speed, process nut fall rate, and chocolate spray rate.

THE PROACTIVE APPROACH

The proactive approach changes the operator's focus. The processes become the driving force for tactical decision-making. With this process focus and using quantitative measures, this truly is statistical process control (SPC). Back at the candy plant, our focus will be on base building, caramel drop, nut sprinkling, and chocolate lay down. The SPC approach requires teamwork from both the strategist and tactician. The team is the operator, procurement and suppliers, process design, product design, marketing, and quality assurance.

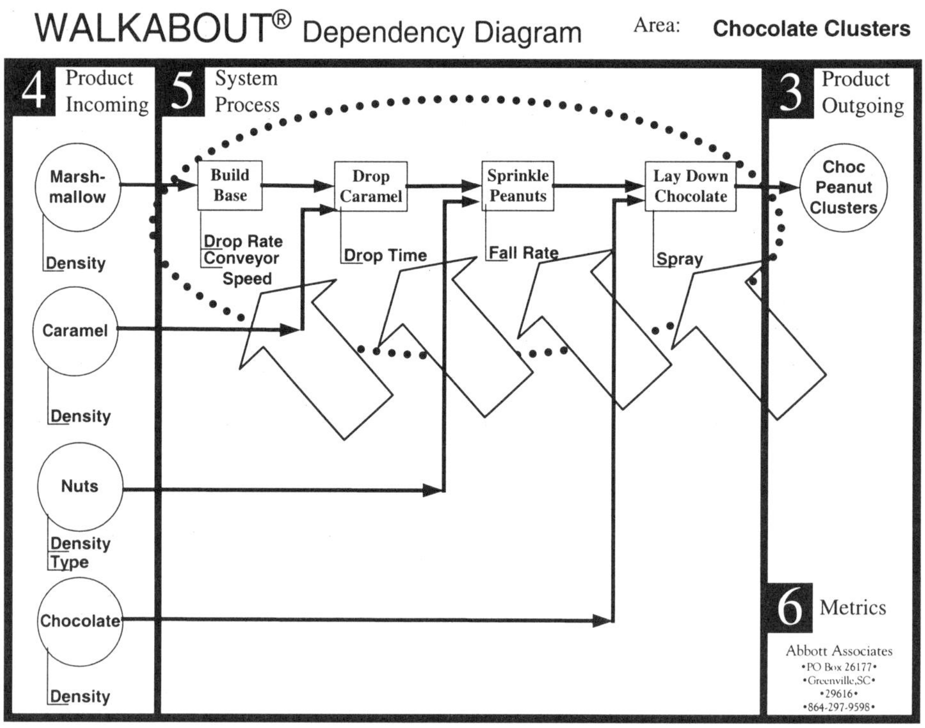

The suppliers and procurement must assure that all incoming products are acceptable. The machine operator is dependent on procurement to supply quality ingredients. Procurement must supply quality nuts with correct density.

In our first step, procurement is supplying quality nuts. Our ingredients are the first component and procurement must be responsible for assuring that all our incoming products are satisfactory. First the incoming peanuts are checked and set with a density of 2.7.

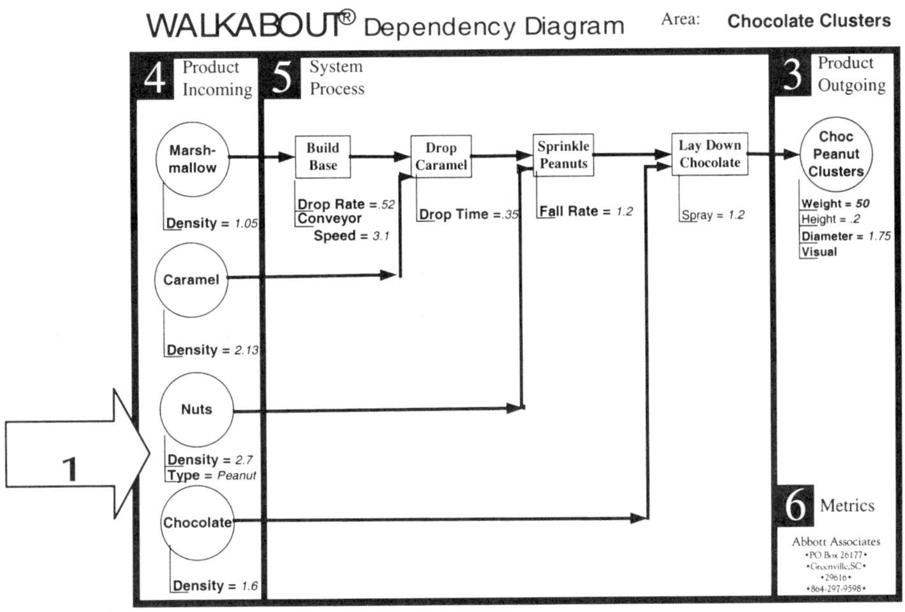

Our tactical operators are dependent on the strategic procurement team member to make sure that we have quality nuts going into our clusters.

The machine and all the incoming product and process settings are then set up. Second, the conveyor speed is set to 3.1. Third the nut fall rate is set to 1.2. Fourth and finally the chocolate spray rate is set to 1.2.

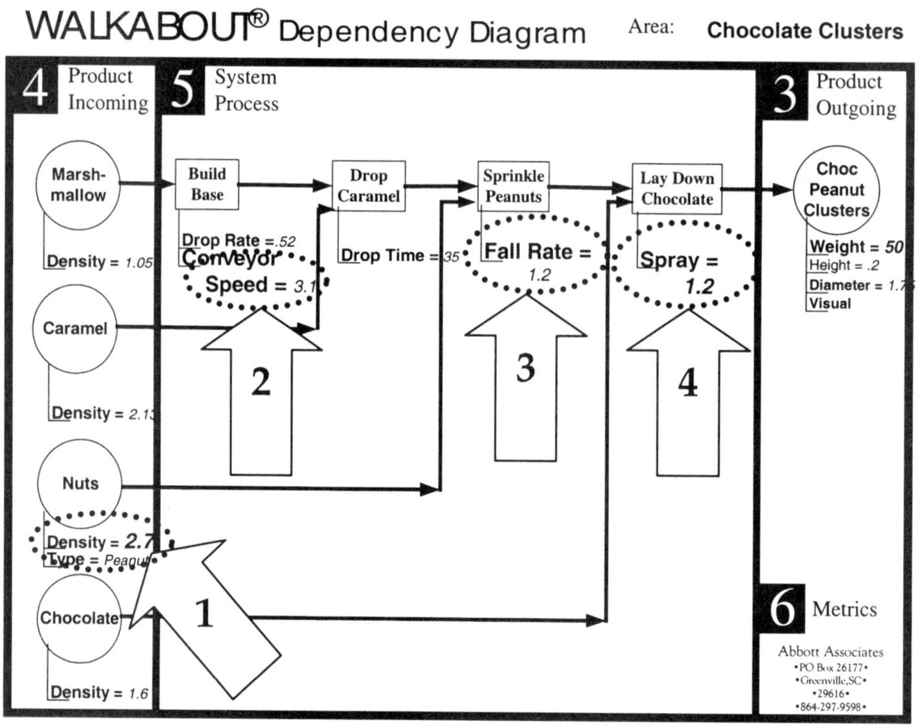

With each process metric set up we check and begin monitoring our chocolate cluster weight, conveyor speed, nut fall rate, and chocolate spray.

SPC adds the element of continuing to measure the process components. After the conveyor speed is set up, measurements of the conveyor speed are taken and then posted to a control chart. Our plotted results for the first day are shown below.

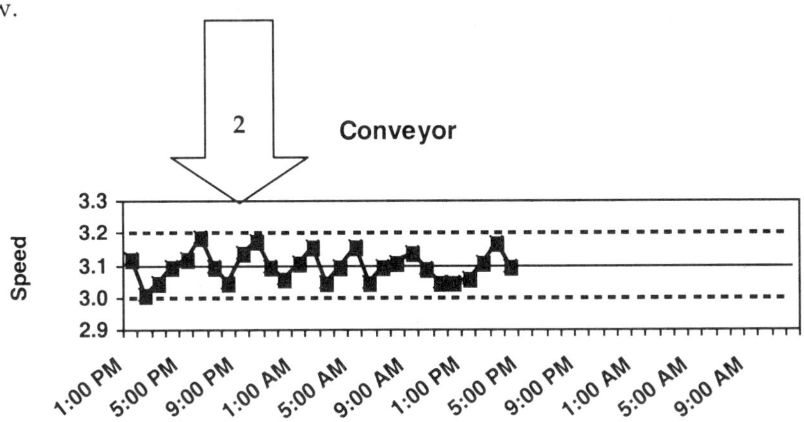

The conveyor speed is moving through the four stations at a consistent speed.

The nut drop rate is set up and once again the operator measures the drop rate on a regular basis. This nut drop rate is posted on the control chart below.

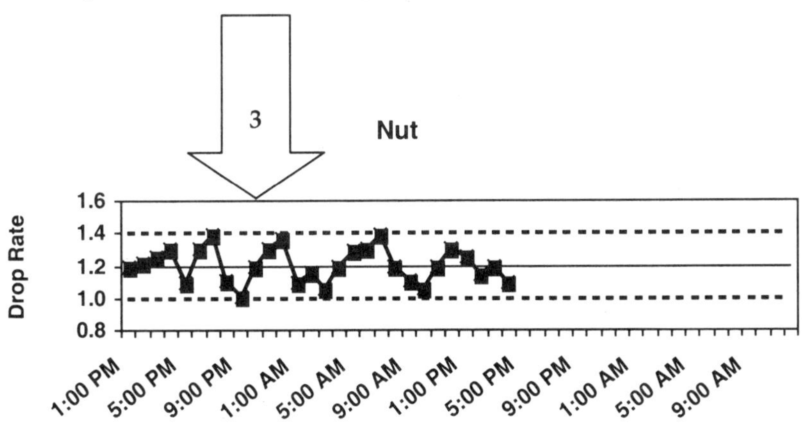

The nut drop rate is also consistent and in control.

Our final process metric of chocolate spray is set up and measurements taken. These chocolate spray measurements are posted on our final control chart.

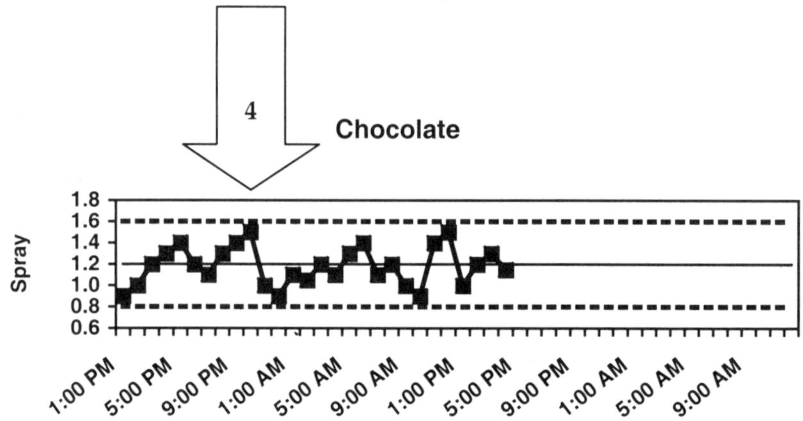

The chocolate spray process metric is also consistent and in control.

With our incoming products tested and all our process metrics setup correctly and monitored for consistency with control charts, we are now ready to monitor our final product. The final product of chocolate clusters can now also be monitored for consistency with its own control chart.

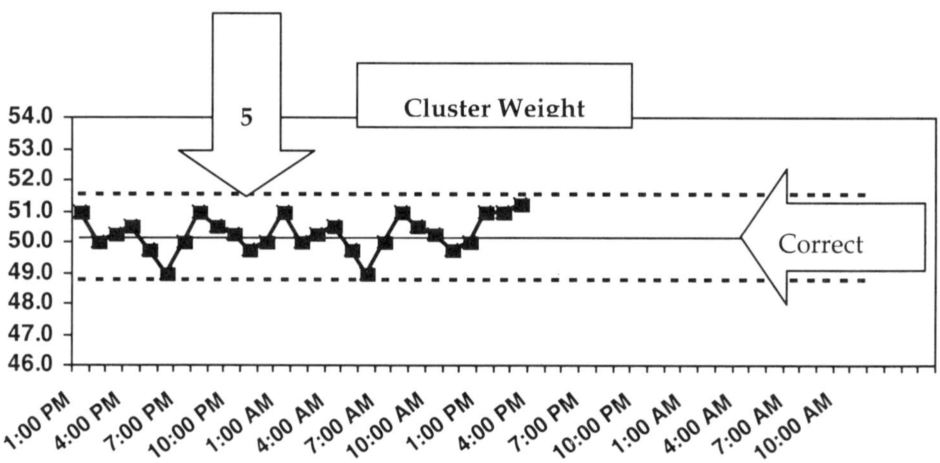

The cluster weight produced from this machine is shown to be consistent.

The operator of the cluster machine can take pride in achieving his role of a correct and consistent machine. This is validated with correct targets for each metric and consistent results documented on the control charts for each metric.

Now the complete picture can be drawn. Our product and processes are in control. The Walkabout® and four control charts show a vivid picture of how our operator is perfectly playing his role for the team. The only thing remaining is a capability study of the chocolate clusters.

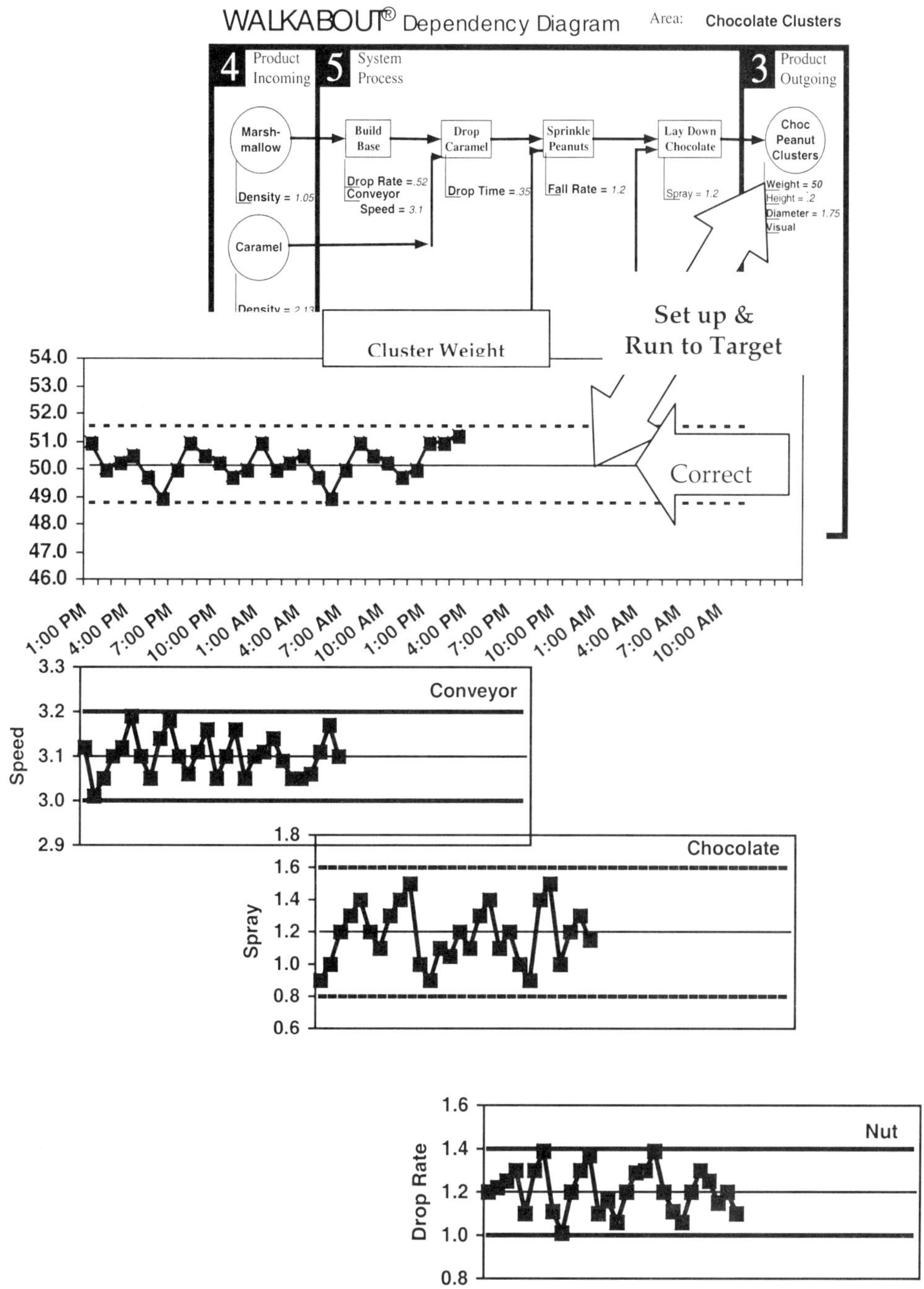

Since the machine is set to the correct target, and is also running consistently for both the product and the process, a capability study is done. The chart below is the graphical results of the capability study.

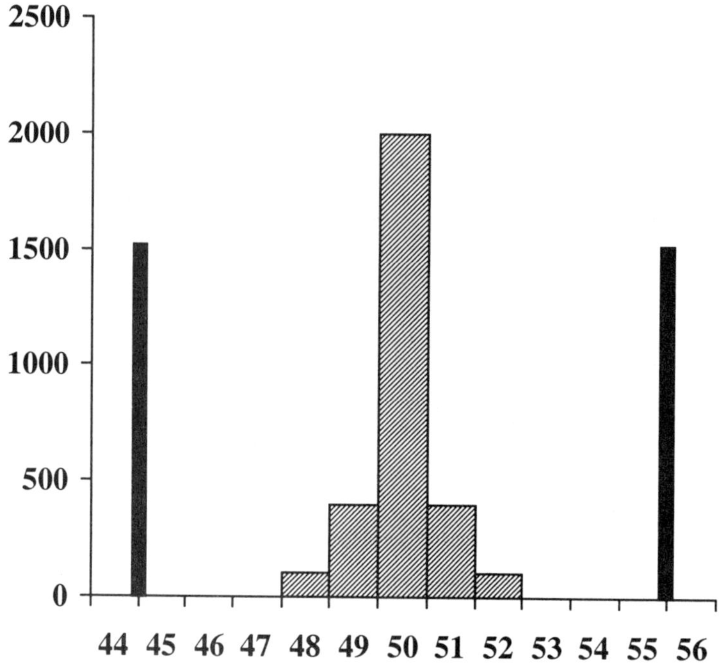

The chocolate cluster machine capability is zero percent defective. As long as the machine is correct and consistent we should continue to expect the product results to be zero defectives.

To continuously monitor both the product and all the process, the operator uses the Walkabout® and four control charts.

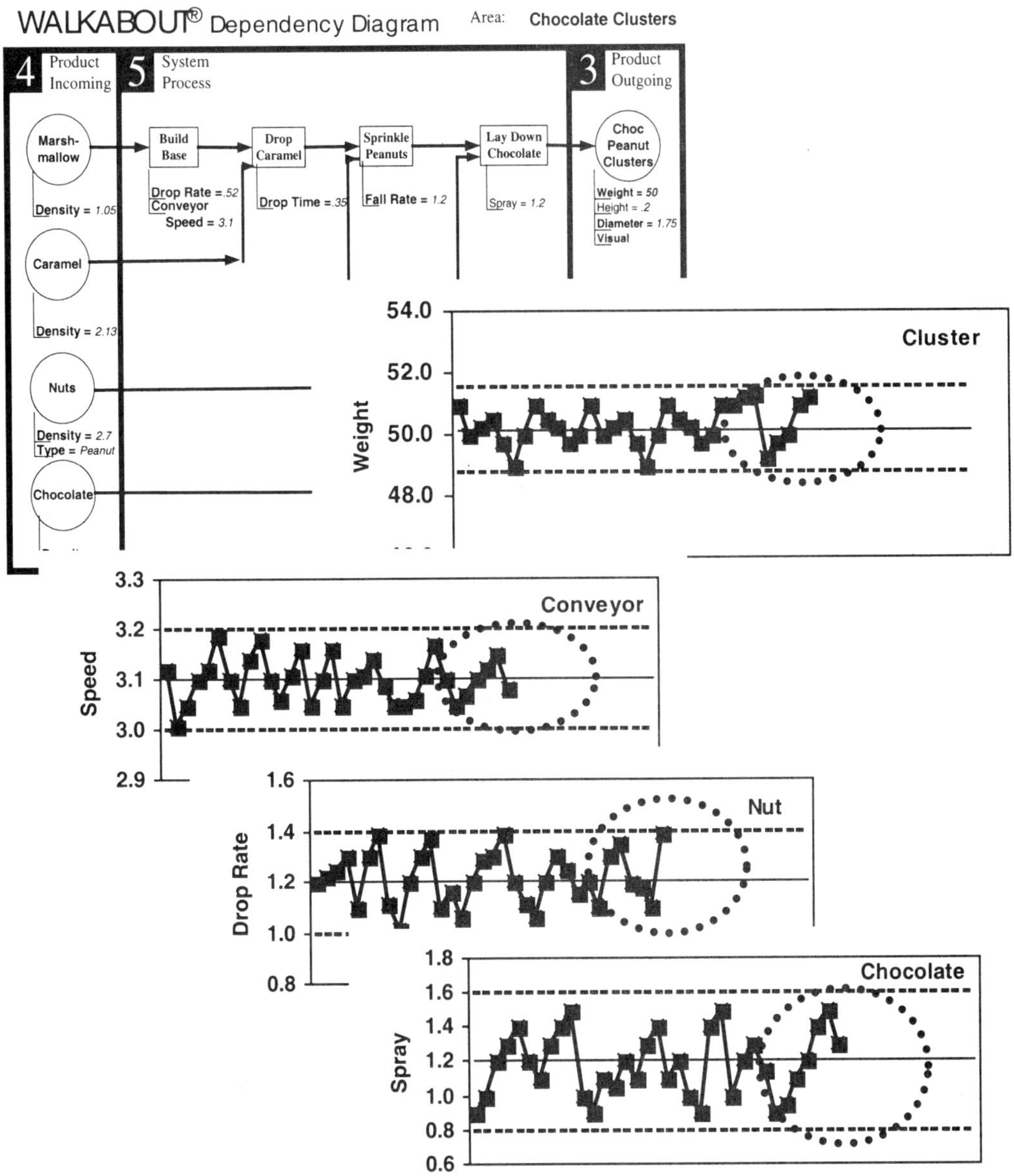

As we move into the second day the circled areas show that both the product metrics and process metrics are still in control. With all metrics consistent as validated by their control charts, we are confident that our capability is zero percent defects.

The monitoring effort continues with the operator checking the four control charts for a change in his machine.

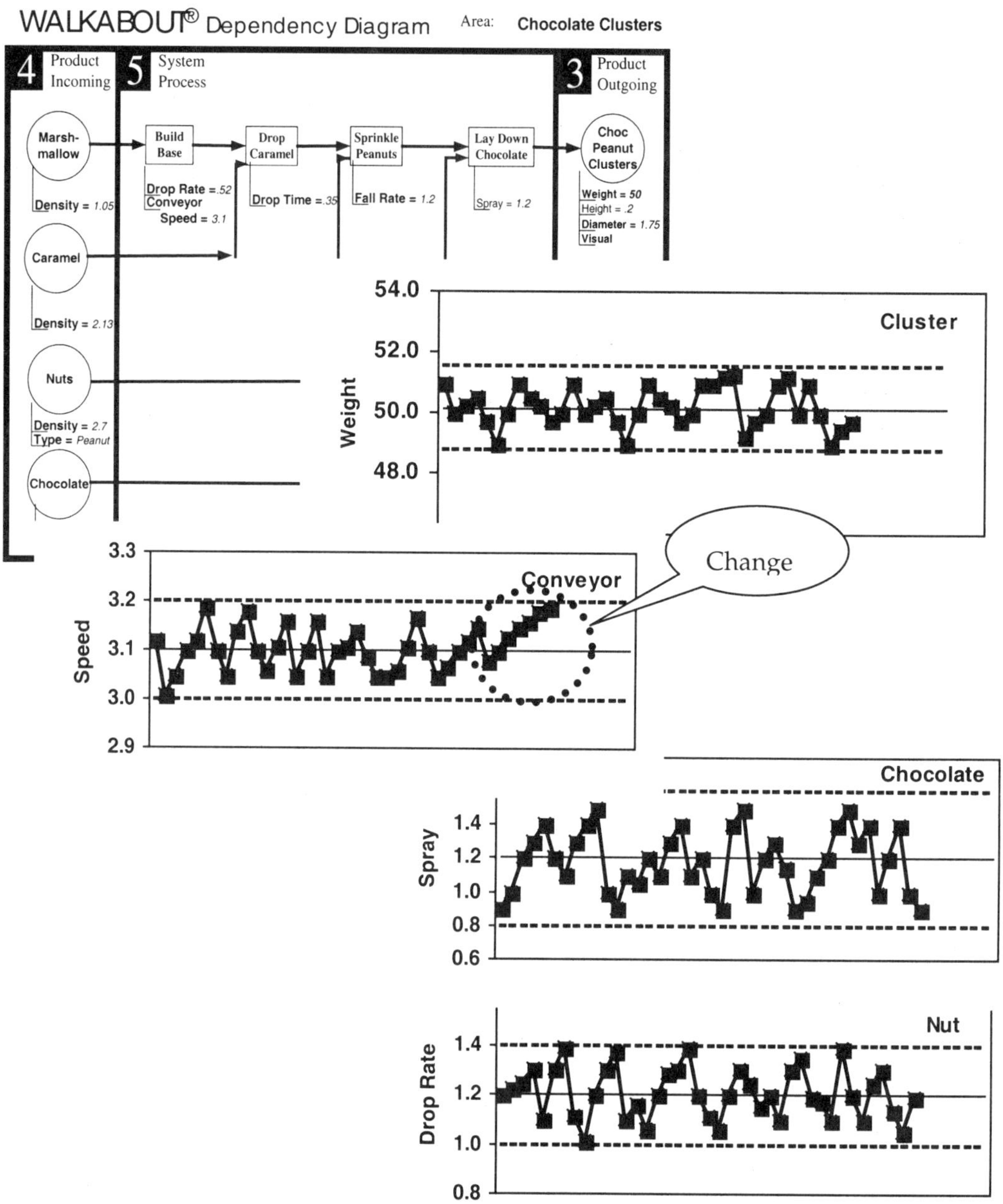

The conveyor speed control chart has violated a control rule. Seven consecutive values, each value progressively higher than the prior, indicates a change.

The operator begins to research and identify the cause of the change. He must also notify the strategic decision-makers of the change so that they can assess the impact of the change on the quality of the product.

Once the strategic process engineer has been notified that a change in the conveyor speed has been detected, we are free to begin our research to find the cause. Since the alarm has occurred on the conveyor speed control chart, the operator's task is simpler. His research focuses on the conveyor speed. The target setting for conveyor speed is 3.1, but the dial on the conveyor has drifted up slightly to 3.2. The actual measurement of the conveyor speed shows a measurement of 3.19. Thus, the dial is calibrated properly to the conveyor speed. With the conveyor going slightly faster than expected, the nuts and chocolate being dispensed are not synchronized.

The strategic decision-maker does a capability study on the cluster product weight to check the impact of the change on the weight. This capability study uses the measurements from the change time period. The pictorial capability study results are shown below.

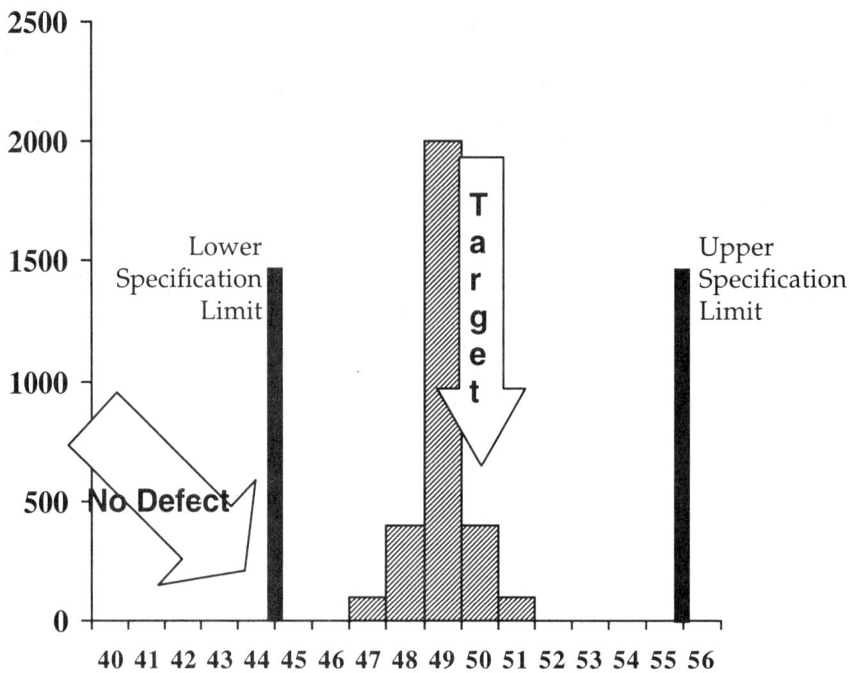

First, no defects have been produced. Our distribution has shifted slightly below the target. This shift is due to the conveyor speed increase identified by the operator's research. By simply resetting the conveyor speed back to the original speed the cluster output should shift back to the target.

All aspects of the operation are improved following the SPC philosophy. The SPC approach is proactive. Problems are fixed when they are small before they have a chance to grow. We make adjustments on the process metrics not on product defects. Changes can be seen sooner. The operator's job is simpler because the control chart changes point him to the metric that has changed. The operator can build and maintain product and process knowledge with this method.

The operator is now ready to make an adjustment confident in the outcome. This confidence is built on the fact-based knowledge that the conveyor speed is faster causing the weight to be lower than target. Even though the weight is lower than perfection, none of the clusters have reached a critical stage where some of the products are defective.

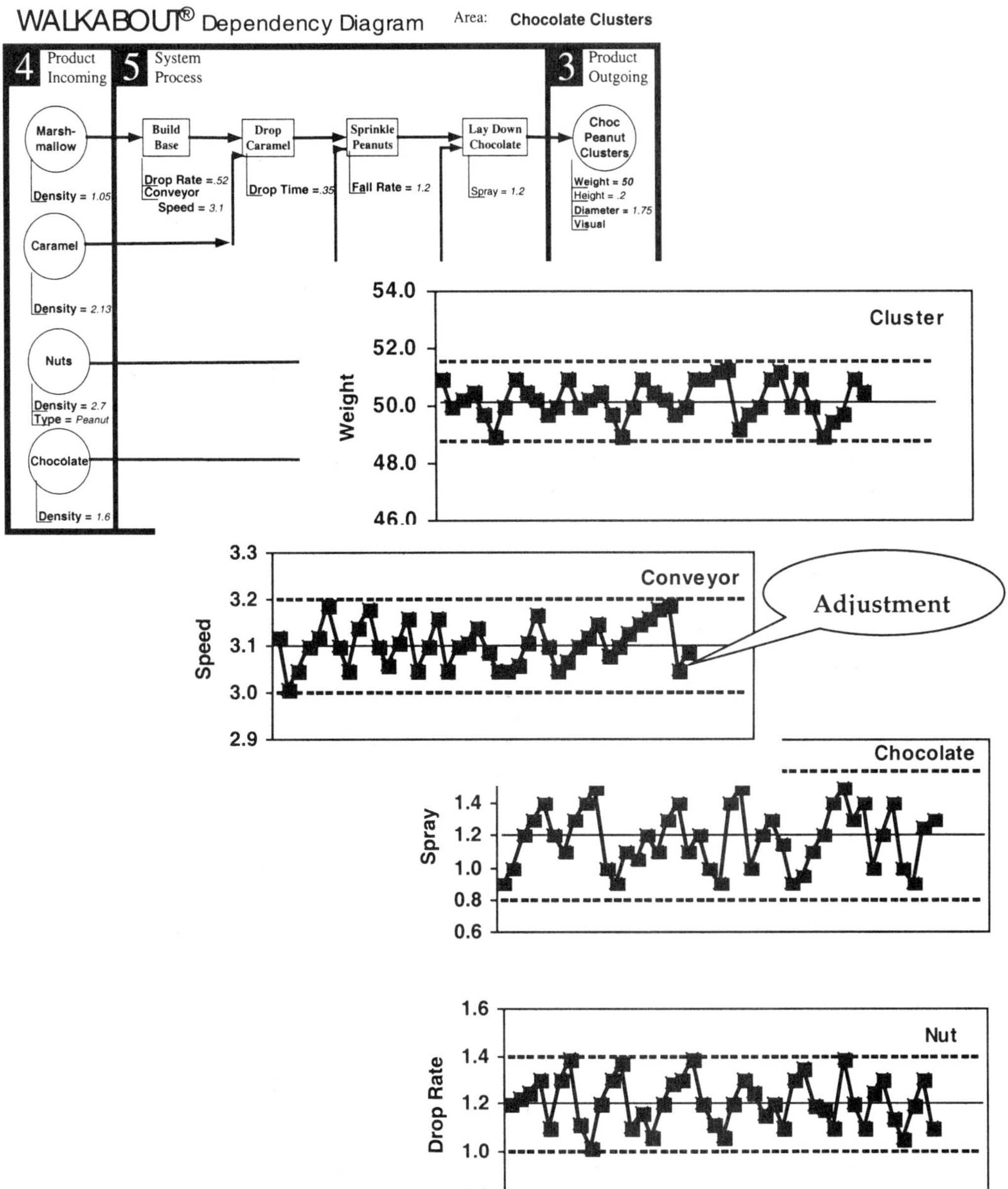

The conveyor speed is adjusted back to its target setting of 3.1. The operator now closely watches all the metrics for change. As the process and product stabilize, the operator returns to his continuous surveillance of the product, process, and their control charts.

THE BEST OF A BAD SITUATION

Now that we understand process control, let's see some of the trials and tribulations of our strategist. First, operations uses the Walkabout® to set up correctly.

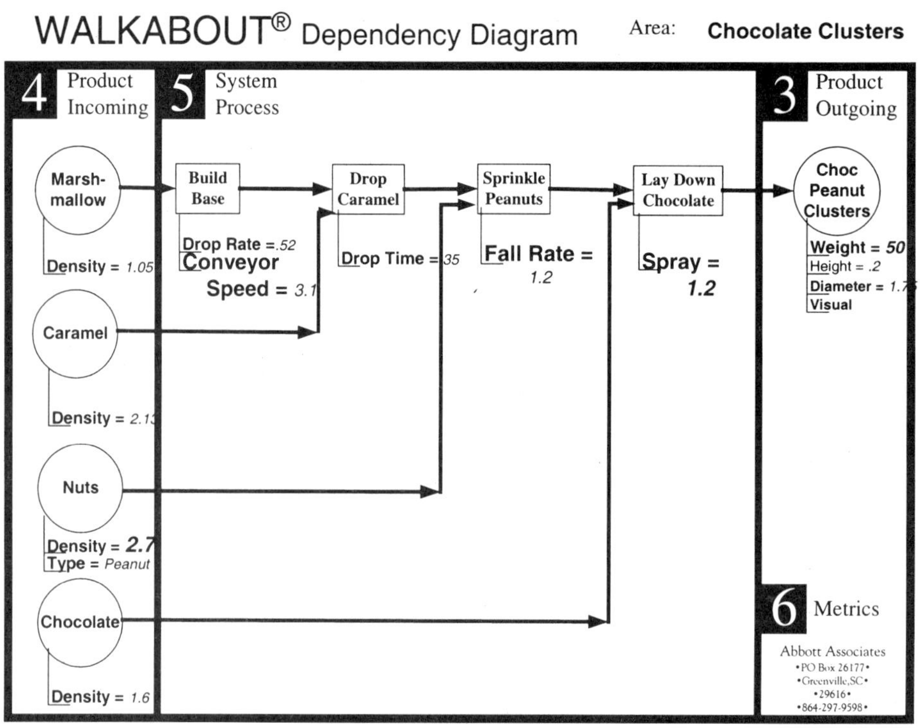

For our example, let's assume all of the operational product and process metrics are correct and consistent at every step of this example.

Because this example deals with strategic issues only the cluster weight control chart will be shown. Our first day's cluster weight results yield very stable results as demonstrated by our control chart below.

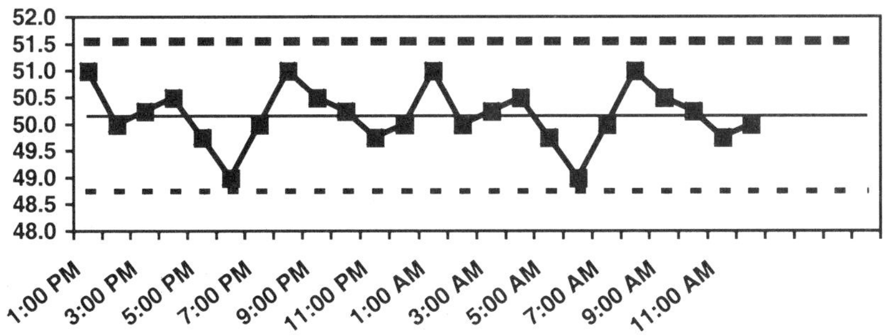

Since the machine is set to the correct target and also is running consistently, a capability study is done. The chart below is the graphical result of the capability study.

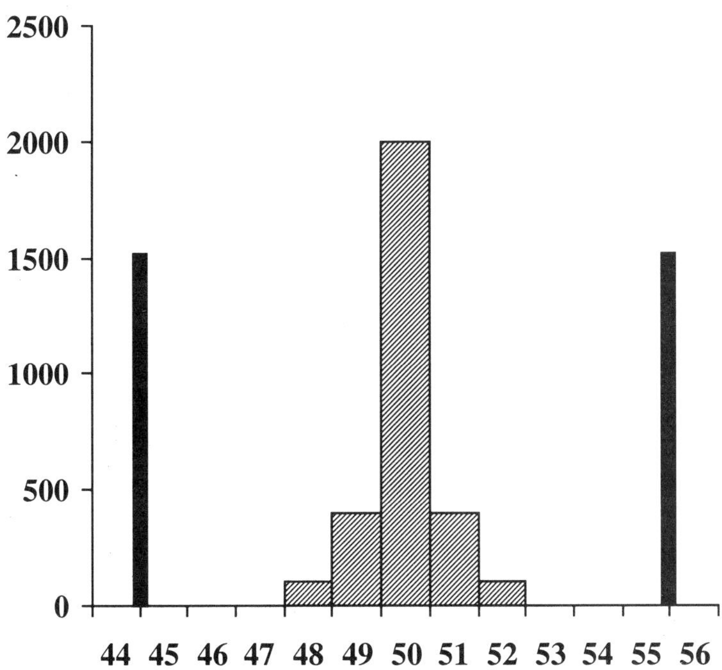

The chocolate cluster machine capability is zero percent defective. As long as the machine is correct and consistent we should continue to expect the product results to be zero defectives.

A new peanut vendor approaches procurement with an opportunity to buy less costly peanuts. Since peanuts are an agricultural product purchases are made for a planting season lasting about six months. The strategic impact and magnitude of this decision is critical. A complete season of bad peanuts can be devastating to our products, but the possibility of a cost savings is enticing.

The cost reduction from this new source is $.010 per chocolate cluster out of the total cost of $.235 per cluster. This would be about a four percent saving per cluster. Our production for a month is 1,200,000 clusters. The savings for this volume is $12,000 per month or $144,000 per year.

This cost reduction is tempting, but product deterioration has to be avoided. As a means of avoiding any problems, a trial is negotiated to test the new supplier and their peanuts. The vendor agrees to supply enough peanuts for a thorough production run.

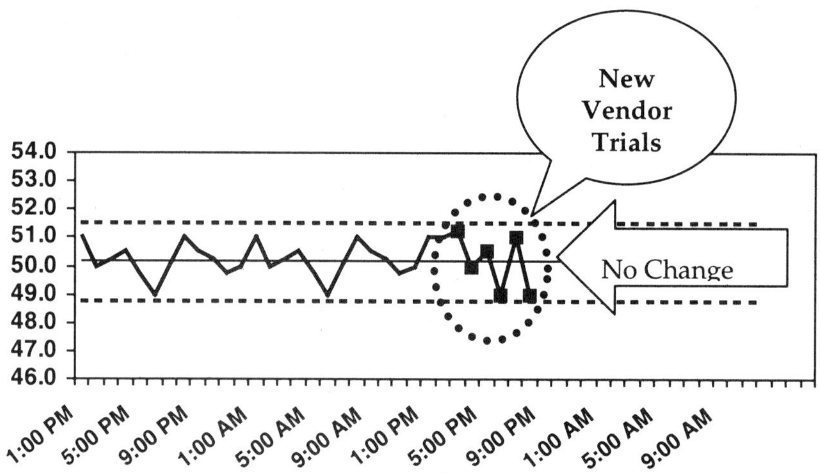

The new vendor peanuts are run in a controlled batch to produce the chocolate clusters. The test procedure has been well thought out, testing all the possible situations that will be encountered in full production.

The clusters using the new peanuts show no change in quality. All the control charts, capability studies, and taste tests show the new, cheaper peanuts give identical results to the previous vendor's peanuts.

Confident that the cheaper peanuts will perform as well as the higher priced peanuts, we place an order with our new supplier for the upcoming season.

Since peanuts are a natural product that are perishable, all the old vendors peanuts are consumed before we start running our new vendor's shipments. This will avoid any losses because of spoilage.

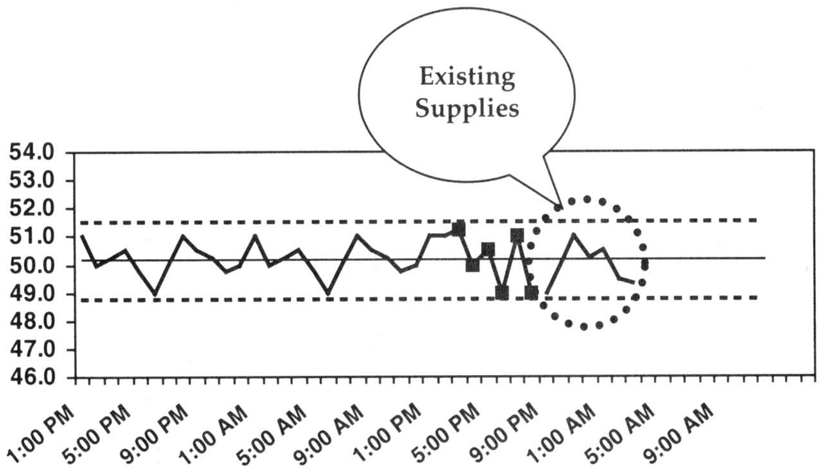

With our supplies of old peanuts depleted, we cut over to our new vendor's peanuts. Immediately our cluster weight control chart signals a change.

Our operators begin the search for the cause while alerting the strategic decision-makers to the change.

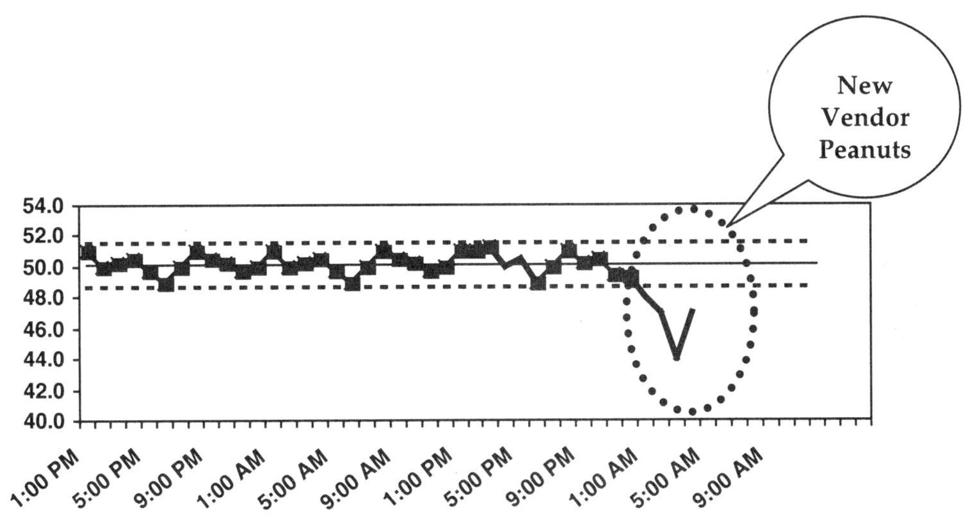

The strategist quickly computes and constructs a capability study. This study shows a defect rate of ten percent.

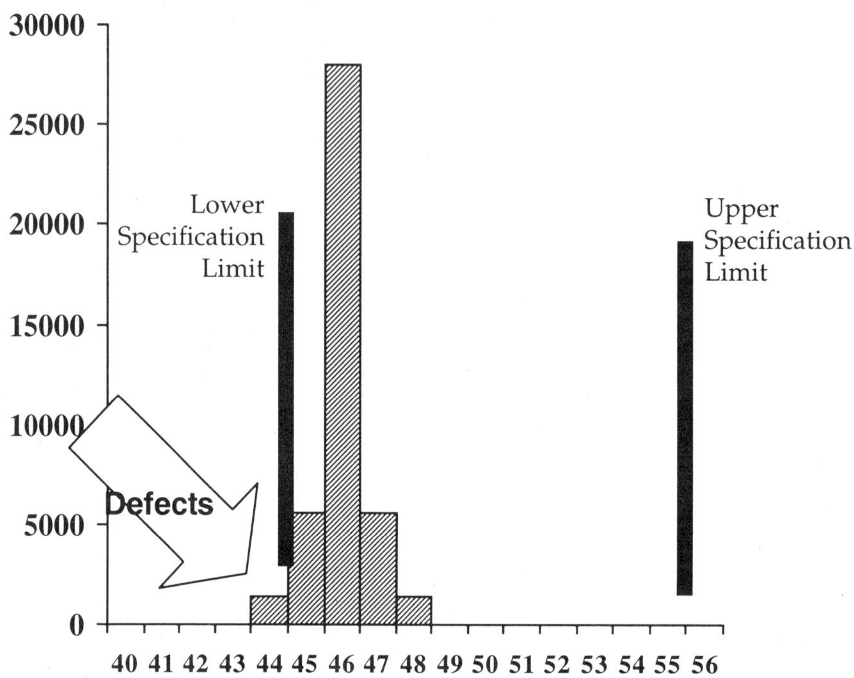

The operators check their processes and find them all correct and consistent. When the new peanuts are reviewed they show up less dense than the old vendor's peanuts and the trial test peanuts.

It's a fine mess they've got us in now. First of all, who is the "they" we are talking about? "They" are the strategic decision-makers. Second, the mess we are talking about is having defective incoming product ingredients. These situations are the reality of everyday life in business – welcome to the real world. The procurement strategist did everything properly in the quest to reduce cost but not jeopardize the product quality, but we still have problems.

Since the strategist got us in this mess, the strategist must get us out of it. Specifically, the tactical operator must not take it upon himself to fix or resolve the problem, nor can the strategist abdicate his role. Let's come up with options and strategies for working our way out of this situation.

The strategist must ultimately decide our course of action. This decision is exclusively the strategist's.

There are three possible strategies. Each option has positive and negative ramifications. The consequences of each option must be identified and considered so that a thoughtful strategic decision can be made.

- Stop running the facility
- Adjust and change the product
- Accept the ten percent defect rate

Each option and its consequences must be discussed in depth. Each option's positives and negatives will be explored in detail.

Stop running the facility

Since our defect rate is ten percent, we must first assure that none of the light chocolate clusters are shipped to the customer. The simplest way to assure this is to stop all production until good peanuts can be obtained. The stoppage will continue until replacement peanuts arrive.

The replacement peanuts can come from our original source, our new source, or a third source. Nuts from our original source or a third source may lose our cost savings and will take a while to begin shipping. With nothing produced no defects will get to the customer. There is a customer down side, in that no clusters are shipped at all. No production, no shipments, no revenue, no profit, leads to no business. No business leads to no jobs. Stopping the facility seems pretty draconian so let's take a look at the next option.

Adjust and change the product

This option is popular but it is not normally thoroughly thought through. If you add extra peanuts or chocolate or caramel or marshmallow to the clusters, it will amount to a totally new product. This new product may or may not be saleable.

This option kills our cost savings and also adds additional cost so that the total cluster cost is higher than when we started.

The decision to choose this option ***must*** be made at a strategic level by marketing, sales, product design, and process design.

Many times tactical personnel have taken this option but failed to notify the strategist. This situation can easily happen if the strategist and tactician are not working as a team. The strategic decision-maker must decide whether or not to select this option.

Accept the ten percent defect rate

This option is accepting the ten percent defect rate while alternative sources of peanuts are identified. Finding the alternative source may take a few days or it may take months. So this strategy amounts to a two-front campaign.

The first front is for the strategic procurement decision-makers to rapidly research the three alternative peanut sources. They must contact the original source, the new source, and any third sources to obtain cost, performance, and delivery commitments to resolve our crisis.

The second front is for operations to continue to make the clusters with the ten percent defects. Even while running this defect rate the tactical focus of running the facility correctly and consistently must be maintained.

While our operators run the facility, all defective products must be removed. For this task we will use a group of quality assurance technicians to remove the defects. This team will weigh every cluster and remove all light clusters. The added testing and high scrap rate will drive our cost up, but by using the quality assurance technicians we avoid a potential problem; the operators will not be tempted to adjust their machines to compensate for the bad clusters.

Strategic Decision Time

None of our options is a good choice but a decision must be made. Our strategic decision-makers choose the best of the bad choices. This is option three, which is to continue to run the facility.

To continue to run the facility, operations must have new values for what now comprises correct and consistent. Correct and consistent is now defined as ten percent defective and ninety percent acceptable clusters. Since the role of operations is to run the provided facility correctly and consistently, the operator must keep his focus on his machine. The strategist must worry about the defect rate and getting a capable facility back in place.

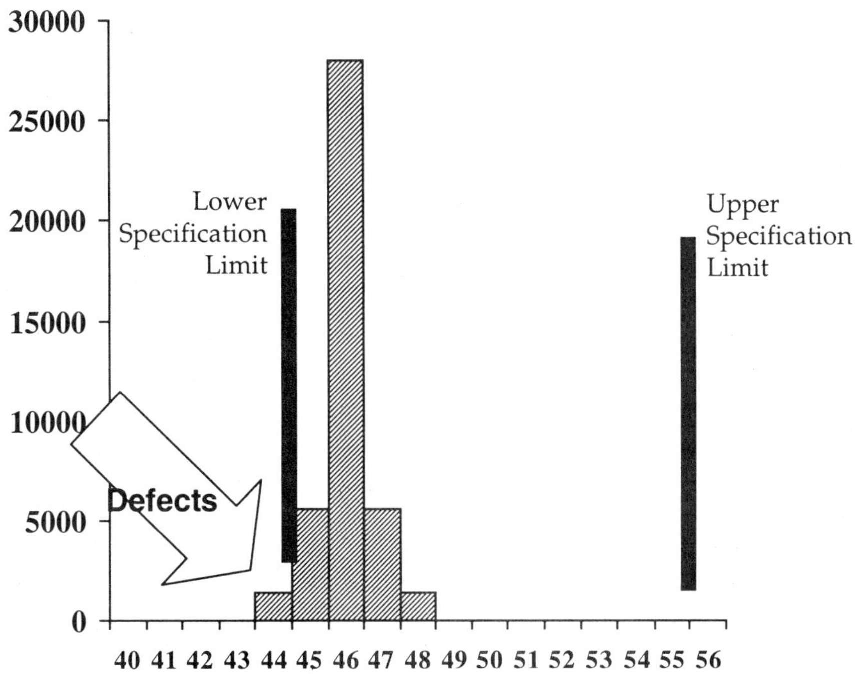

The operator is provided with a revised Walkabout®, cluster target weight, and new control limits on the control chart.

Correctness must be revised on the Walkabout® to have a new cluster target weight of 46 grams.

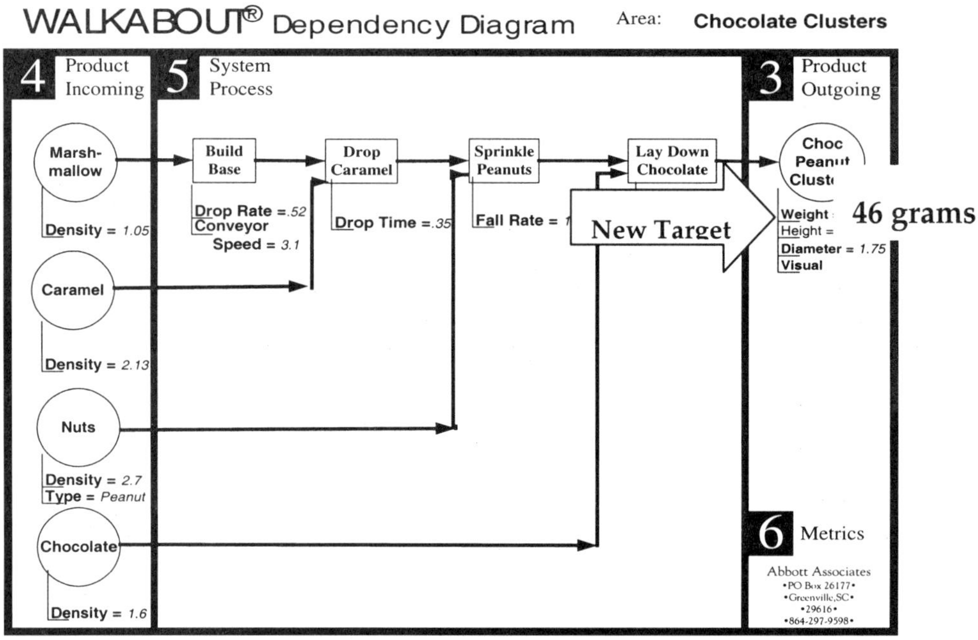

The new control limits are shown below. The control chart is now reflecting the ten percent defect rate.

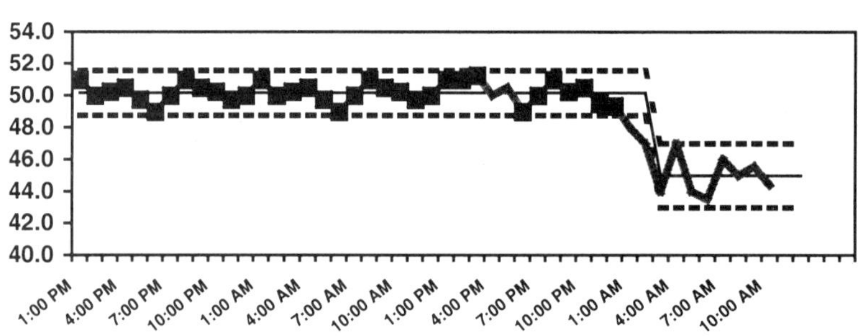

Do not think these new operator tools imply that we are happy with our defect rate. Only fixing the facility can change this defect rate.

A staff is provided to perform a 100% testing program. Each cluster will be weighed and all light clusters (weighing less than 45 grams) will be removed and destroyed. Only those clusters weighing more than 45 grams will be accepted and shipped to customers.

Once the strategist has obtained an agreement with our old supplier, and we return to using their peanuts, the change should trigger the control chart alarm.

The strategist checks our capability again.

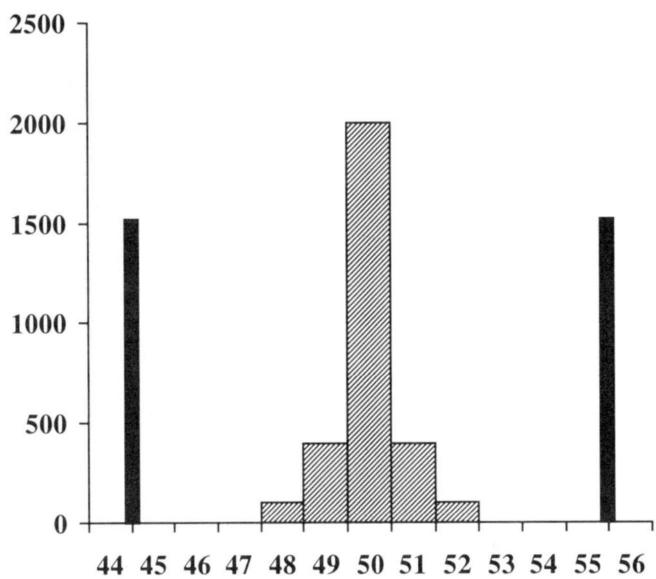

With our capability returned to zero percent defects and our old supplier peanuts back, the control limits are again revised to reflect a state of consistency.

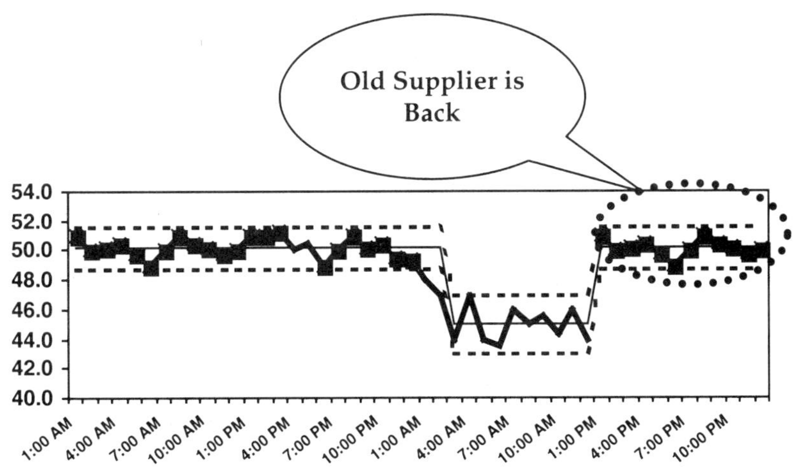

Our target cluster weight on the Walkabout® is returned to 50 grams.

Remember, correct and consistent (as described in a control chart) is directly tied to the level of actual capability. This removes the crutch that operations can't do its job because the strategist has not supplied a capable machine or facility. With Division of Labor as our guide, our tactical labor force must keep their focus on correctly and consistently running the facility no matter how good or bad the facility. Correctly and consistently running the provided facility is a tactical issue. Capability is a strategic issue. Tacticians and strategists must work together with brilliance to achieve the best performance from each machine.

Every in-coming product, process, and out-going product metric must be tracked, measured, monitored, and analyzed. Our tactician must keep their focus on the correct and consistent process, trusting that the strategists do their job of providing the capable facility.

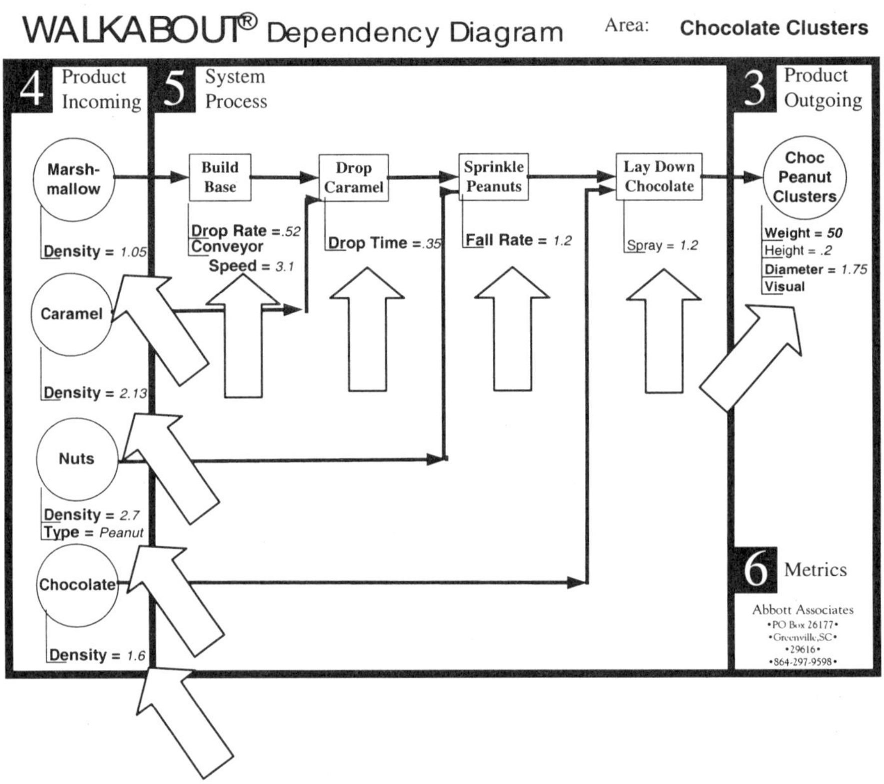

Chapter Nine:

Benefits of Running your Machines with SPC

WALKABOUT® BASE-CAMP

Starting in the front of this book, we have built the case for running a business, or an improvement process, like an expedition ascending a mountain to its summit. The mountain climbing expedition's essentials include having the right equipment, access to the essentials of life (food, water, air), having a stable launch point, and properly trained team members. These essentials are referred to as the base-camp for a mountain climbing expedition, and in our improvement process, a base-camp will also be required. The operational base-camp's components are correctness, consistency, and capability forming the Walkabout® Base-camp.

RESULTS OF RUNNING YOUR MACHINES WITH SPC

By reaching the Walkabout® Base-camp, our minimum expectation will start with *superior products. Products that are as close to our strategic design targets,* as our machines are capable of achieving.

In addition, more productivity, less scrap, and less rework will *reduce unit cost.*

With every machine treated as unique, our strategic decision-makers will understand the ability of each machine. With this knowledge they can *detect and repeat improvements, replicate improvements to other machines, fix non-capable machines, and stop problems before they create defects, added cost, or missed deliveries.*

HOW WILL RESULTS BE ACHIEVED?

The total organization works as a team because both strategic and tactical players understand their role. The tactical operators focus their attention on the correct and consistent running of their machine. Running each machine correctly and consistently and to its maximum capability will yield the desired results of better performance at lower cost with timely delivery.

Each machine is run to its maximum design potential. This maximization is due to clearly defined roles and methods, brilliant execution with a process focus, and a clear understanding of the impact. When the machine is run correctly and consistently, the knowledge from the machine's capability study will facilitate optimum performance of each machine.

With correct methods defined, documented, and executed, process setup will be faster.

Quickly responding to change stops problems while they are small. This quick response allows us to identify the cause of any improvements so that they are continued and replicated into the future.

Let's give a quick recap of the Walkabout® Base-Camp and the tools that support it.

WALKABOUT® BASE-CAMP MILESTONES

Milestone One - Correctness

We, the tactical workforce, are dependent on our strategic decision-makers. First, the strategic correctness must be established. A correct, functional organization with clearly defined roles must be established to cover marketing and sales, product design, process design, manufacturing, and quality assurance. The product and process factorial families must be correctly calculated. A clearly defined product and a detailed method must be created. The strategic decision-makers must clearly communicate these correct methods. The strategist will provide a Walkabout® diagram of the method. On each metric a target will be provided.

WALKABOUT® Dependency Diagram Area: **Chocolate Clusters**

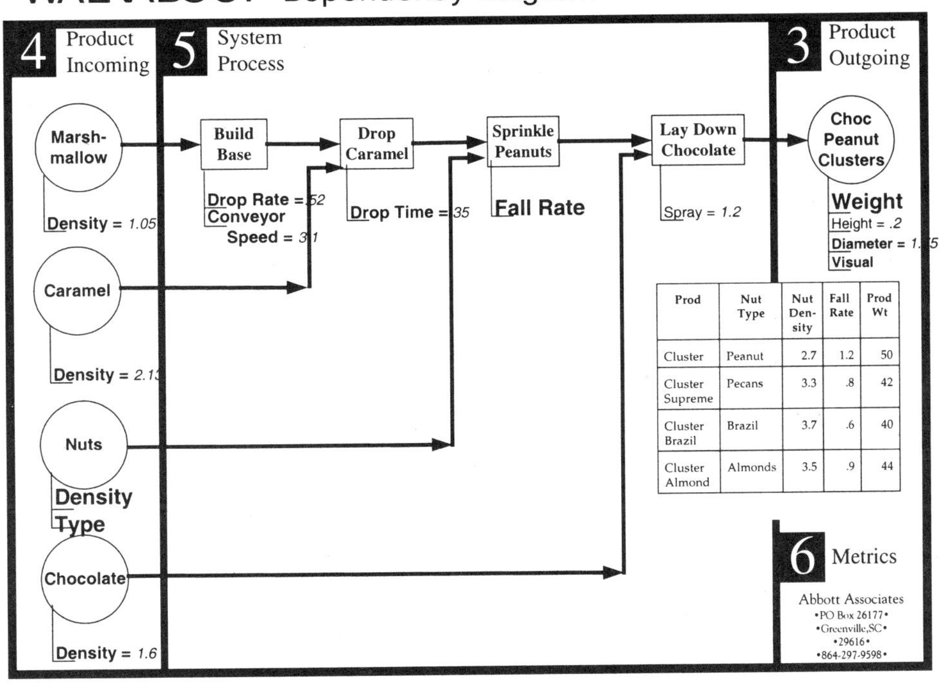

Prod	Nut Type	Nut Den-sity	Fall Rate	Prod Wt
Cluster	Peanut	2.7	1.2	50
Cluster Supreme	Pecans	3.3	.8	42
Cluster Brazil	Brazil	3.7	.6	40
Cluster Almond	Almonds	3.5	.9	44

Correct also includes focusing on the processes. The tactical workforce will use the proactive approach of SPC. With a clear understanding of their role, the machine operators can focus on the process metrics. These metrics and quick response to change allow a preventive approach to problems.

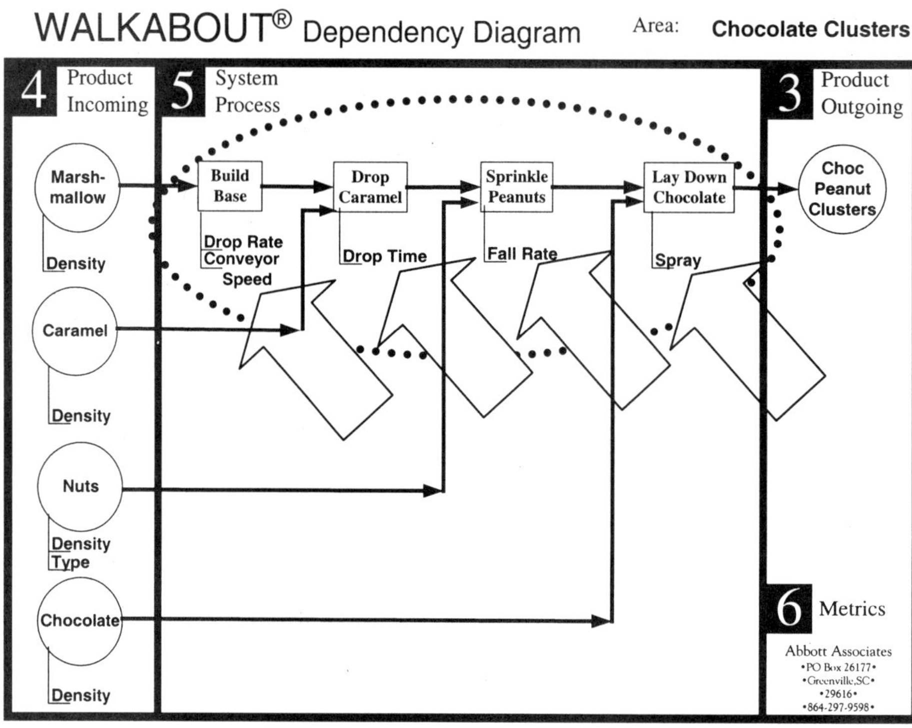

The tactical workforce must brilliantly execute the plan, methods, and the target settings. This brilliant execution should not be confused with perfection. As human beings, we will always strive for perfection but the reality is that it may be outside our grasp. Keep trying to correctly execute the process, but don't worry too much about being absolutely perfect.

Milestone Two - Consistency

The second step toward our base-camp requires monitoring each metric for every machine to determine how consistent each product and process metric is. This consistency provides a way of predicting what will be produced time after time. Each machine must have its own consistency established. This consistency is monitored through the use of a mathematical tool called control charts.

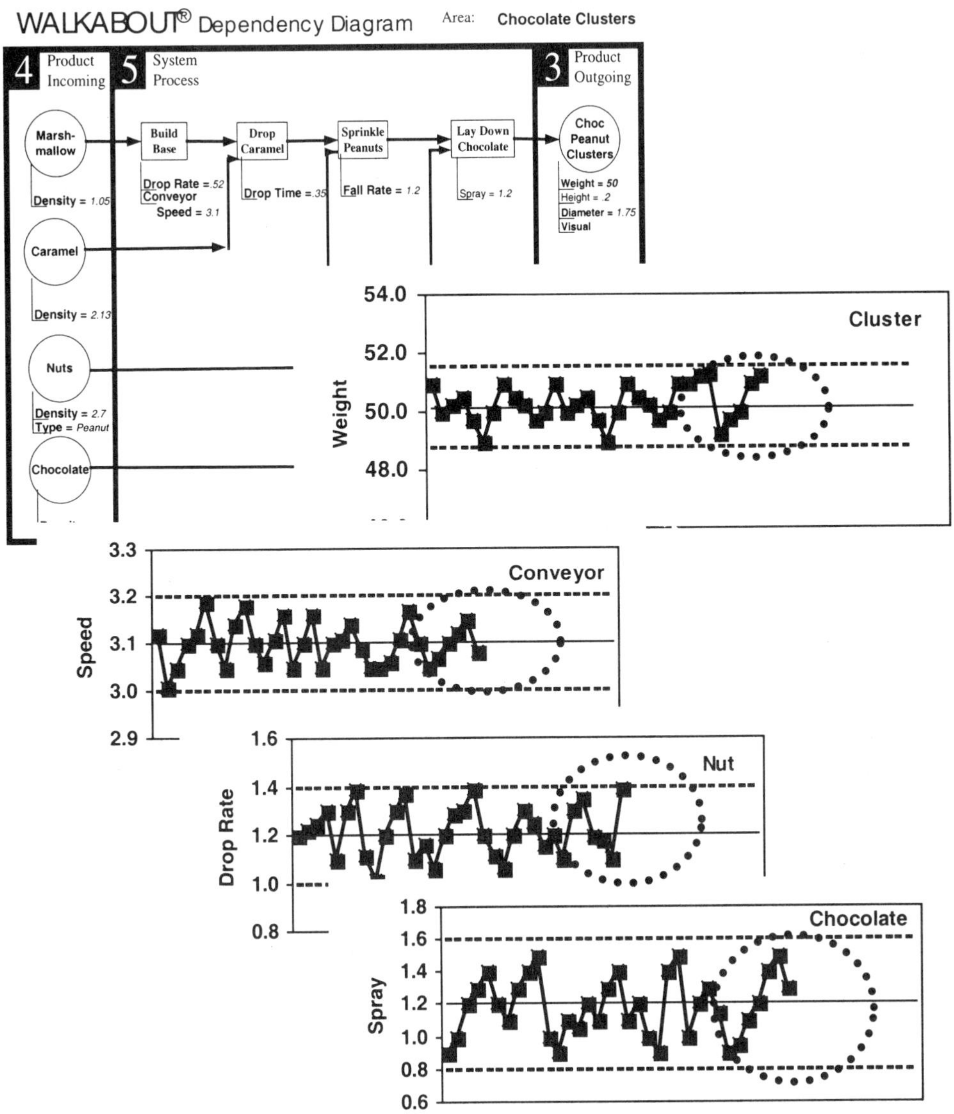

Once the tactical product and process issues of correct and consistent are accomplished, our strategic teammates can become active.

Milestone Three - Capability

Once the state of correctly and consistently run machines or processes is reached, the strategic decision-maker is positioned to assess how well the product we make meets the customer's expectation. The strategist uses a capability study like the one below.

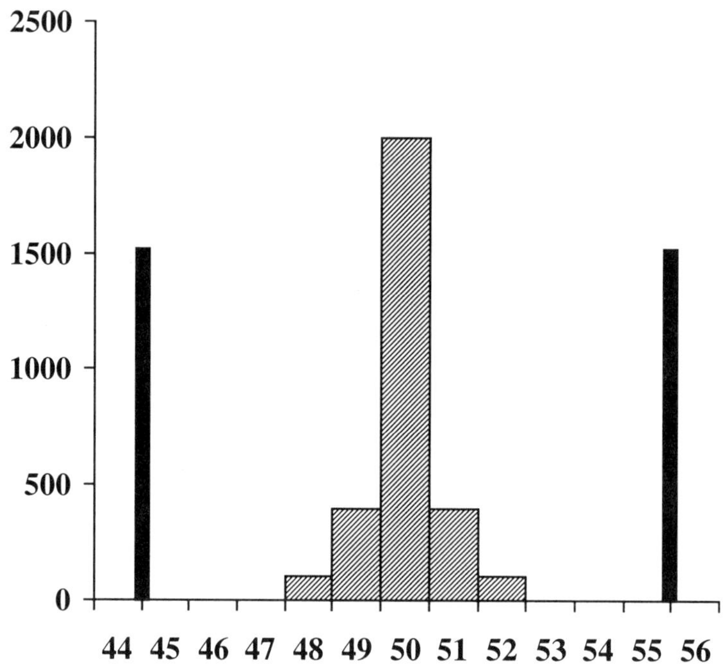

Take special note: correctly and consistently running the provided machine are tactical issues. Capability is a strategic issue. Both the tactician and strategist must work together with brilliance to achieve the best performance from each machine.

When the Walkabout® Base-camp is complete, we begin to reap the fruits of our labors. The three Cs (correct, consistent, and capable) of the Walkabout® Base-camp are the backbone of a profitable business.

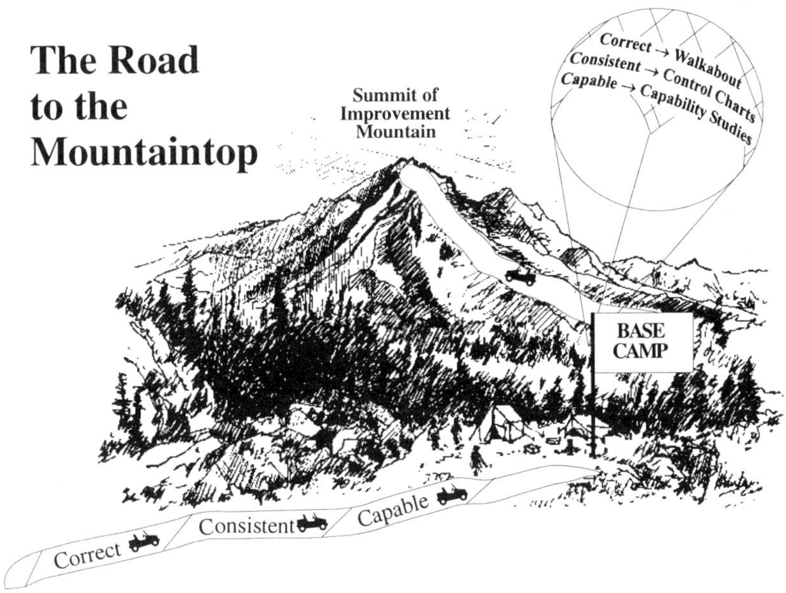

THE BIG PICTURE

The goal of *Running your Machines with SPC* is to run each machine to its maximum potential all the time. All our efforts are marshaled toward reaching that goal. Running each machine to its maximum potential is vital to an effective operation.

The bigger objective is satisfied customers and stockholders. This objective can be defined as delighting our customer with products that meet or exceed their expectations, while reducing our production unit cost, and providing our products to our customers the moment they want them. The three components of this bigger objective are:

- **Better quality products (performance)**
- **Reducing our cost (cost)**
- **Delivery when the customer wants the product (time)**

The performance, cost, and time criteria define the bigger objective.

With the tools of the Walkabout® Base-camp in place, reaping the results is a matter of execution and focus. Achieving improved performance, reducing our cost, and better delivery to our customers is now possible, and will continue.

Appendix A: Control Chart Calculations

$\overline{X}$ and R Charts

A_2 is the conversion factor to arrive at three (3) standard errors for use in determining the upper control limit (UCL) and lower control limit (LCL) of $\overline{X}$ chart.

D_3 is the conversion factor to arrive at three (3) standard errors for use in determining the lower control limit (LCL) of R chart.

D_4 is the conversion factor to arrive at three (3) standard errors for use in determining upper control limit (UCL) of R chart.

The table below is used to compute our control limits and is based on the use of the number of samples in the subgroup.

Subgroup Size	A_2	d_2	D_3	D_4
2	1.880	1.128		3.267
3	1.023	1.693		2.574
4	0.729	2.059		2.282
5	0.577	2.326		2.114
6	0.483	2.534		2.004
7	0.419	2.704	0.076	1.924
8	0.373	2.847	0.136	1.864
9	0.337	2.970	0.184	1.816
10	0.308	3.078	0.223	1.777
11	0.285	3.173	0.256	1.744
12	0.266	3.258	0.283	1.717
13	0.249	3.336	0.307	1.693
14	0.235	3.407	0.328	1.672
15	0.223	3.472	0.347	1.653

$\overline{X}$ and R CHART Conversion

$\overline{X}$ Chart Control Limit Calculations

$$UCL_{Ave} = \overline{\overline{X}} + A_2 * \overline{R}$$

$$LCL_{Ave} = \overline{\overline{X}} - A_2 * \overline{R}$$

R Chart Control Limit Calculations

$$UCL_R = D_4 * \overline{R}$$

$$LCL_R = D_3 * \overline{R}$$

Remember, all control charts are tactical tools to detect change in the product and process. A control chart can never tell you how well the product meets the customer's expectation.

$\overline{X}$ and S Charts

A_3 is the conversion factor to arrive at three (3) standard errors for use in determining UCL and LCL of $\overline{X}$ chart where $A_3 * \overline{S}$ is an approximation of three (3) standard errors.

B_3 is the conversion factor for the approximation of three (3) standard errors used as the LCL on the S chart.

B_4 is the conversion factor for the approximation of three (3) standard errors used as the UCL limit on the S chart.

Subgroup Size	**A_3**	**C_4**	**B_3**	**B_4**
10	0.975	0.9727	0.284	1.716
11	0.927	0.9754	0.321	1.679
12	0.886	0.9776	0.354	1.646
13	0.850	0.9794	0.382	1.618
14	0.817	0.9810	0.406	1.594
15	0.789	0.9823	0.428	1.572
16	.763	.9835	.448	1.552
17	.739	.9845	.466	1.534
18	.718	.9854	.482	1.518
19	.698	.9862	.497	1.503
20	.680	.9869	.510	1.490
21	.663	.9876	.523	1.477
22	.647	.9882	.534	1.466
23	.633	.9887	.545	1.455
24	.619	.9892	.555	1.445
25	.606	.9896	.565	1.435

$\overline{X}$ **and S CHART Conversion**

$\overline{X}$ chart control limits

$$UCL_{Ave} = \overline{\overline{X}} + A_3 * \overline{S}$$

$$LCL_{Ave} + \overline{\overline{X}} - A_3 * \overline{S}$$

S chart control limits

$$UCL_S = B_4 * \overline{S}$$

$$LCL_S = B_3 * \overline{S}$$

Individual chart

E_2 is the conversion factor to arrive at three (3) standard errors for use in determining UCL and LCL of the Individual chart.

D_3 is the conversion factor to arrive at three (3) standard errors for use in determining LCL of the R chart.

D_4 is the conversion factor to arrive at three (3) standard errors for use in determining UCL of the R chart.

Subgroup Freq	E_2	d_2	D_3	D_4
2	2.660	1.128	-	3.267
3	1.772	1.693	-	2.574
4	1.457	2.059	-	2.282
5	1.290	2.326	-	2.114
6	1.184	2.534	-	2.004
7	1.109	2.704	0.076	1.924
8	1.054	2.847	0.136	1.864
9	1.010	2.970	0.184	1.816
10	0.975	3.078	0.223	1.777

Individual Chart Conversion

$\overline{X}$ chart control limits

$$UCL_{Ave} = \overline{X} + E_2 * \overline{R}$$

$$LCL_{Ave} = \overline{X} - E_2 * \overline{R}$$

Moving Range control limits

$$UCL_R = D_4 * \overline{R}$$

$$LCL_R = D_3 * \overline{R}$$

Remember, all the variable charts use the same interpretation rules.

$\overline{X}$ and R Control Chart Form

Area & Machine: _________ Metric: _____________

Date: _________

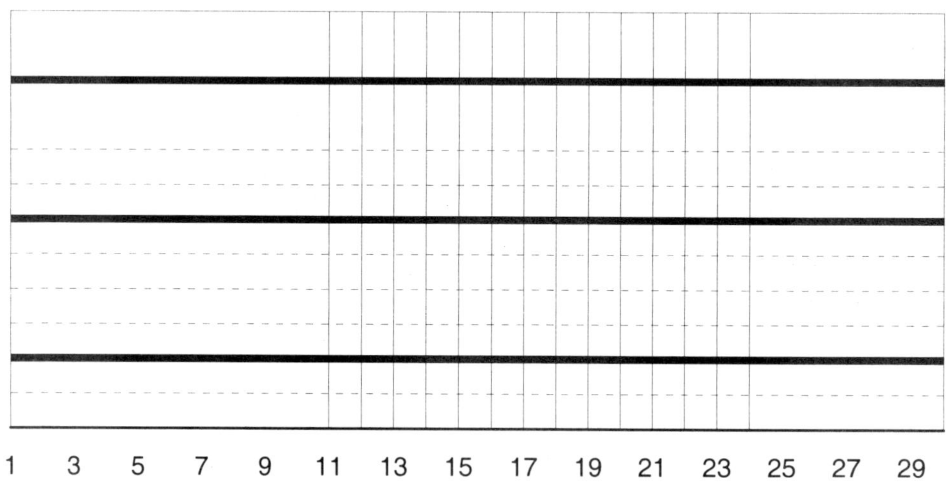

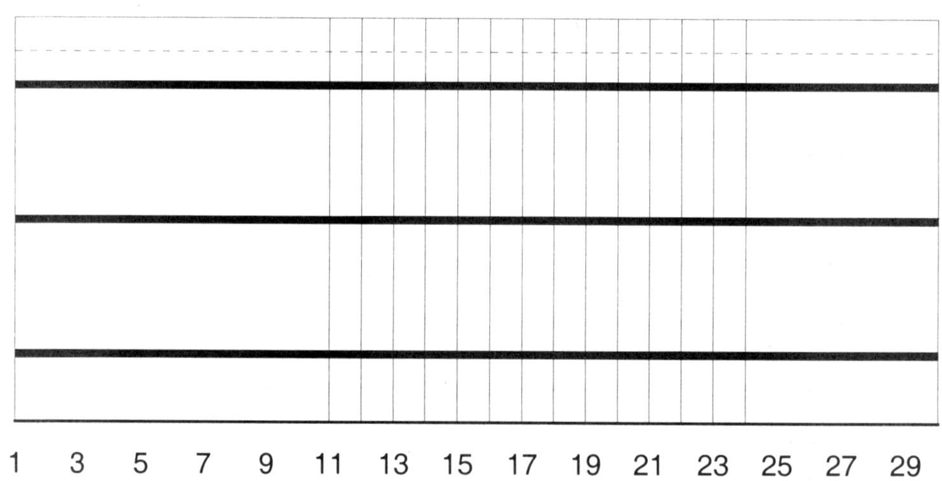

Subgroup Size	A_2	d_2	D_3	D_4
4	0.729	2.059		2.282
5	0.577	2.326		2.114
6	0.483	2.534		2.004

$\overline{X}$ and R CHART Conversion

$\overline{X}$ and S Control Chart Form

Area & Machine: _________ Metric: _____________

Date: _________

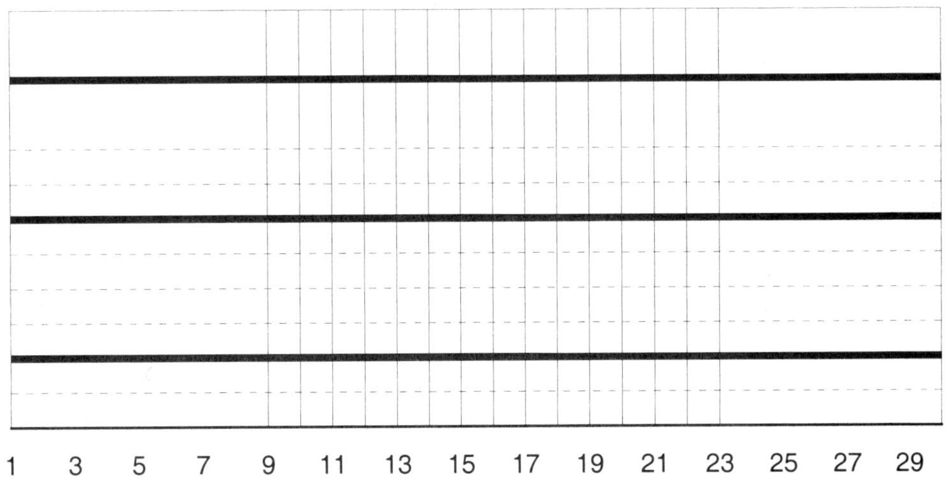

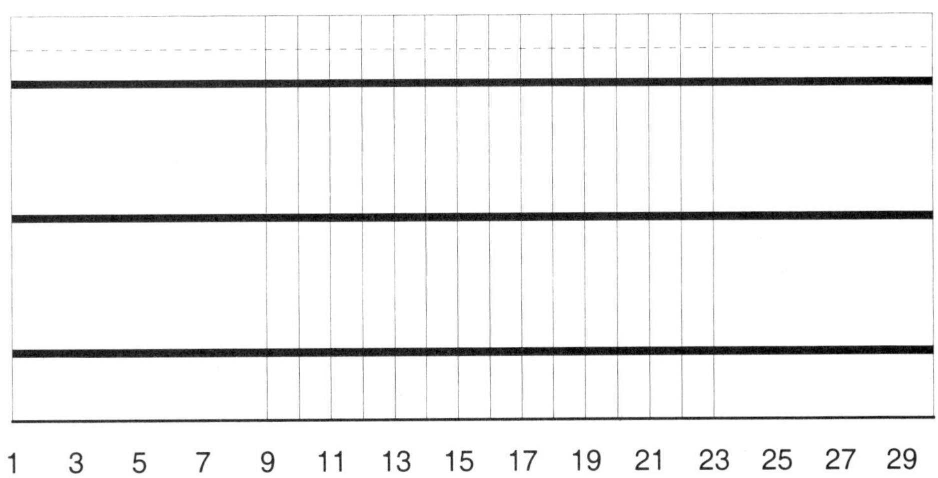

Subgroup Size	A_3	C_4	B_3	B_4
10	0.975	0.9727	0.284	1.716
11	0.927	0.9754	0.321	1.679
12	0.886	0.9776	0.354	1.646

$\overline{X}$ and S CHART Conversion

Individual Control Chart Form

Area & Machine: _________ Metric: _____________

Date: _________

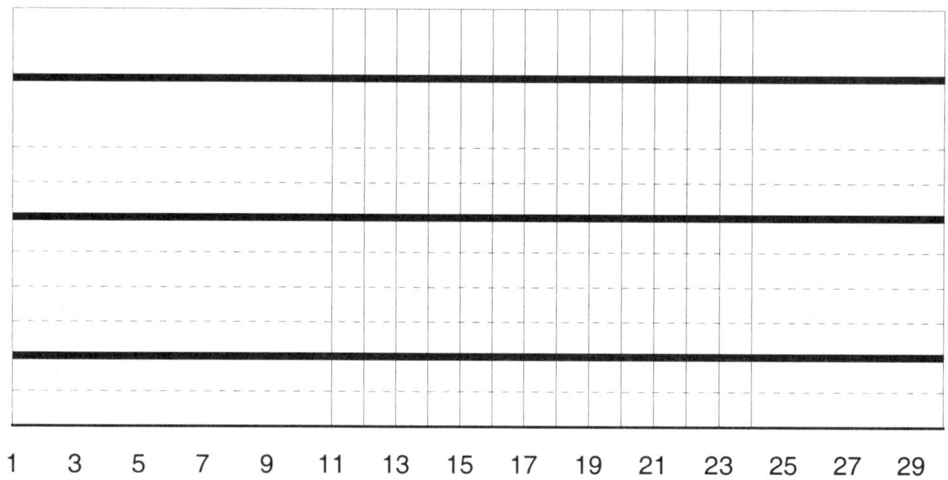

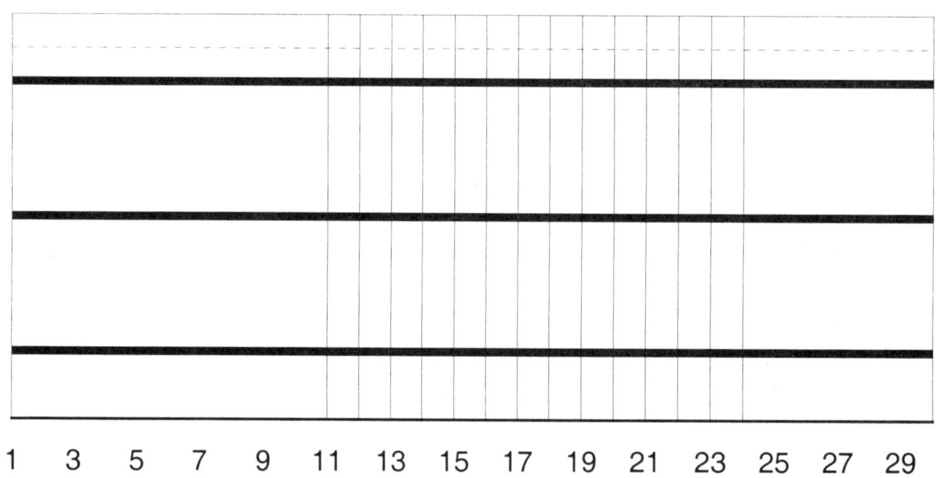

Subgroup Freq	E_2	d_2	D_3	D_4
2	2.660	1.128	-	3.267
3	1.772	1.693	-	2.574

Individual Chart Conversion

Appendix B:
Plotting Control Chart Data

EXERCISES FOR $\overline{X}$ AND R CONTROL CHARTS

The calculated $\overline{X}$ and R-values are listed in the table below.

	1:00	2:00	3:00	4:00	5:00	6:00	7:00	8:00	9:00	10:00
	50	55	60	45	40	50	55	60	45	40
	51	56	61	46	41	51	56	61	46	41
	52	57	62	47	42	52	57	62	47	42
	49	54	59	44	39	49	54	59	44	39
	48	53	58	43	38	48	53	58	43	38
$\overline{X}$	*50*	*55*	*60*	*45*	*40*	*50*	*55*	*60*	*45*	*40*
R	*4*	*4*	*4*	*4*	*4*	*4*	*4*	*4*	*4*	*4*

This is what the $\overline{X}$ data points look like when plotted on a control chart.

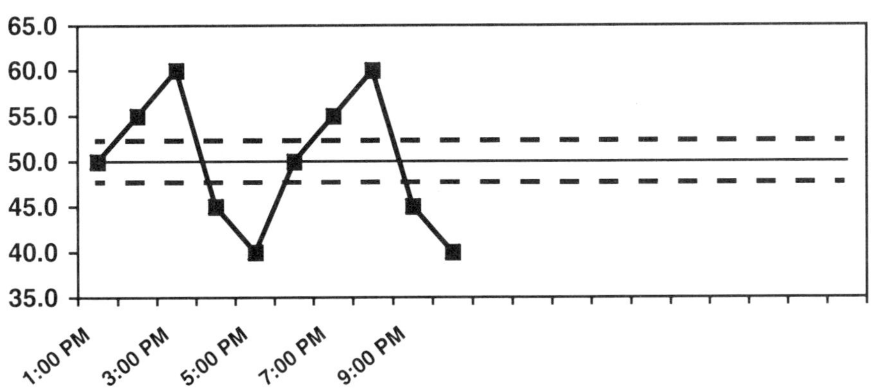

This chart has eight values outside the control limits, and is clearly out of control.

This is what the R-values look like plotted on a control chart.

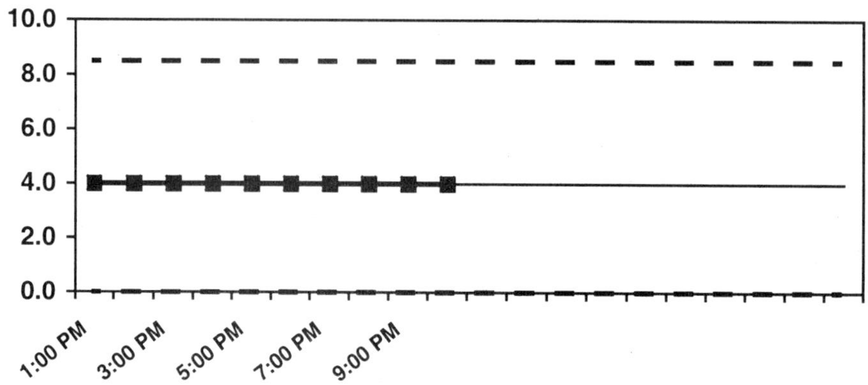

All of the values are clustered around the center. Being clustered around the center is also a control chart alarm.

Appendix C: Glossary

Alarm	A signal that warns or alerts.
Assignable causes	Those causes which definitely result in a particular effect.
Attribute control chart	A control chart that detects a change in discrete data.
Attribute data	Data that is characterized as counts or percentages.
Base-camp	The correct, consistent, and capable running of a facility, along with the information that allows us to achieve these goals.
Benchmark	A standard to compare a value against.
Capability indices	Strategic tools for monitoring capability.
Capability study	The strategic assessment of the facility's ability to make the product the customer wants.
Capable/capability	The strategic ability of a machine, process, or product to meet the customer's expectations.
Categorical data	See ***Attribute Data***
Causes	Why something happened or changed.
Common causes	Causes outside one's control.
Consistent	A uniform and stable running of our operation.
Continuous data	See ***Variable Data***
Control	A mechanism to guide an operation through the detection of change. Consistency is a synonym for control.
Control chart	The tactical graphing tool used to detect changes in a product or process.
Control limits	A tactical tool for operators to detect change in the usual running of a machine.
Correct	The proper running of an operation.
Cost	The amount of time, material, and resources required to make the product.
Discrete data	See ***Attribute Data***
Distribution	The shape of the spread of a set of data points.

Effects	The result or consequences of a cause.
Execution/execute	The carrying out of a task.
Grouping / groups	**Physical:** composed of items that are materially the same. Members of physical groups share characteristics such as type, size, color, same machine or operator. **Associational:** made up of things that are similar but not necessarily exactly the same. Members of associational groups are also of like kind. **Time-series:** composed of material that is similar over time. Members of time-series groups are manufactured at a time when consistency is demonstrated.
In control	The condition when a process is consistent and stable in its natural state.
Individual chart	A control chart for detecting change of a single reading.
Knowledge base	The product and process information gathered to run an operation.
LCL	The designation for the Lower Control Limits.
Lower Control Limits	The lower control limit is three standard errors below the mean value. No measurements should fall below this limit unless a change has occurred.
Lower Specification Limits	The lower limit where the product is deemed unacceptable.
Management	The strategic decision-makers for an organization.
Metric	A study area
Moving Range chart	A control chart that detects a change in variability of an individual set of values.
Nominal value	The target value for a metric.
Normal distribution	The symmetrical, bell shaped curve.
Operations	The tactical decision-makers for an organization.
Out of control	The situation when a change has been detected.
Performance	The product's ability to meet the customer expectations.
Principles of Process Management	The basic truths required to run an operation.
Process	Activities that build the product.
Product	The result that we are trying to manufacture.

Production	Product that has been shipped and paid for.
Quality	Meeting or exceeding the customer's perceived expectations and requirements while reducing cost and providing the product to the customer when he wants it.
Quality Hell	The Good Intentions but flawed efforts at making a quality product.
R chart	The control chart designed to detect movement or changes in the variability of the process.
Special causes	Causes within one's control.
Specification limits	Points at which customers will consider the product unusable.
Statistical Process Control (SPC)	A Statistical Quality Control tool to monitor a continuous operation. Statistical Process Control is comprised of control charts and capability studies.
Statistical Quality Control (SQC)	The set of numeric tools to monitor and maintain a characteristic of a process or product.
Statistics	Facts or data of a numerical kind, assembled, classified, and tabulated so as to present significant information about a given subject.
Strategy/Strategic	The art of achieving a goal. This term is used in both decision-making and goal setting. For more information on this term, additional reading can be found in *Optimize Your Operation*, by James C. Abbott.
Subgroup	Sample results for a particular timeframe.
Tactics/Tactical	The art of achieving a goal utilizing the *available means.* This term is used in both decision-making and goal setting. For more information on this term, additional reading can be found in *Optimize Your Operation* by James C. Abbott.
Tampering causes	Changes that are caused by too much control. Tampering occurs when we are trying to take action on something that should not be changed.
Target	The amount or point at which the customer considers the product to be perfect. The strategic decision-makers should explicitly define the target.
Time	Making sure the product is delivered to the customer when he wants it.

Tolerances	The amount of incompetence that will be accepted by the customer. The acceptable amount of leeway from the target. The tolerances are used to compute the specification limits.
UCL	The designation for the Upper Control Limit.
Upper Control Limit	Three (3) standard errors above the mean value. No measurements should fall above this limit unless a change has occurred.
Upper Specification Limit	The upper limits where the product is deemed defective.
USL	The designation for the Upper Specification Limit.
Value	A measurement for a metric.
Variability	The amount of spread in a set of measurements.
Variable data	Data that is characterized by dimensional readings like length, weight, diameter, etc.
Walkabout® Dependency Diagram	A tool to communicate the correct method and targets for running a machine. A metric blueprint for studying, analyzing, and sharing product and process knowledge between all associates.
Walkabout® Method	A plan complete with clear roles, tools, techniques, and methods designed to correctly, consistently, and capably run an operation.

Index

Robert Houston Smith Publishers titles are available at your favorite bookshop, or online at **www.RHSPublishers.com**.

Title	No. of copies		Price/ each	Subtotal
SPC, third edition		x	$49.95	$
SPC Workbook for the third edition		x	$19.95	$
Optimize Your Operation		x	$39.95	$
Running Your Machines with SPC		x	$29.95	$
			Subtotal	$
			5% tax (SC residents only)	$
			Shipping & Handling (Cont. US: $6 for every two books)	$
			TOTAL	**$**

Ship to:

Name:

Company:

Street/ Physical Address:

City, State, Zip:

Bill to:

Name:

Company:

PO Box/ Mail Address:

City, State, Zip:

Return this completed form to:

Robert Houston Smith Publishers · PO Box 2193 · Easley, SC 29641-2193
Tel: 888-747-6484 · Fax: 888-747-3291
Email: Books@RHSPublishers.com

Robert Houston Smith Publishers titles are available at your favorite bookshop, or online at **www.RHSPublishers.com**.

Title	No. of copies		Price/ each	Subtotal
SPC, third edition		x	$49.95	$
SPC Workbook for the third edition		x	$19.95	$
Optimize Your Operation		x	$39.95	$
Running Your Machines with SPC		x	$29.95	$
			Subtotal	$
			5% tax (SC residents only)	$
			Shipping & Handling (Cont. US: $6 for every two books)	$
			TOTAL	$

Ship to:

Name:

Company:

Street/ Physical Address:

City, State, Zip:

Bill to:

Name:

Company:

PO Box/ Mail Address:

City, State, Zip:

Return this completed form to:

Robert Houston Smith Publishers · PO Box 2193 · Easley, SC 29641-2193
Tel: 888-747-6484 · Fax: 888-747-3291
Email: Books@RHSPublishers.com

For more information about how Methods and Management Consulting can improve your bottom line or for a list of classes (including SPC), contact the team at Abbott Associates Incorporated.

Abbott Associates Inc.
PO Box 26177
Greenville, SC 29616
Tel: 864-297-9598
Fax: 864-297-8624
Email: Info@AbbottWalkabout.com
www.AbbottWalkabout.com